LAB MANUAL TO ACCOMPANY

THE STANDARD VERSION OF

STARTING OUT WITH C++

THIRD EDITION

by Tony Gaddis

Dean DeFino

Salisbury University

Michael Bardzell

Salisbury University

Scott/Jones Inc.
P.O. Box 696,
El Granada, California 94018
Voice: **650-726-2436**
Facsimile: **650-726-4693**
E-mail: **marketing@scottjonespub.com**
Web page: //**www.scottjonespub.com**

ISBN 1-57676-051-0

Lab Manual to Accompany the Standard Version of Starting Out with C++, Third Edition, by Tony Gaddis
Dean DeFino & Michael Bardzell

ISBN: 1-57676-051-0

0 1 2 3 Z Y X

The authors wish to acknowledge, with gratitude, Dr. David Parker (Salisbury University) and Judy Walters (North Central College) for their contribution to the development of this book.

The publisher wishes to acknowledge the memory and influence of James F. Leisy. Thanks, Jim. We miss you.

Text Design and Composition: Cecelia G. Morales
Cover Design: Stephen Adams
Book Manufacturing: Von Hoffmann Graphics

Scott/Jones Publishing Company
Publisher: Richard Jones
Editorial Group: Richard Jones, Mike Needham, Denise Simon, Michelle Windell, and Patricia Miyaki
Production Management: Leslie Austin and Heather Bennett
Marketing and Sales: Victoria Judy, Hazel Dunlap, Donna Cross, Page Mead, and Michelle Windell
Business Operations: Michelle Robelet, Cathy Glenn, and Natascha Hoffmeyer

Contents

Lab Manual Introduction

To the Student. . .

Closed laboratories in computer programming is a vital activity for helping students gain valuable programming skills under the guidance of an instructor. Many different opinions concerning the content of such labs have been generated over the past few years, ranging from programming assignments to scheduled exercises using prepared materials. Although this manual emphasizes the latter approach and has pre-developed code for the students to complete or edit, there are assignments that ask students to independently create small programs which can be assigned as lab activities for students who finish early or as post-lab homework. These student generated code assignments are not intended as a substitute for larger programming assignments. They are small programs designed to test the student in the material given in most lessons.

The length of the lab activities vary from fifty minutes to two hours, depending on the particular institution. For this reason, the manual is divided into "Lesson sets," each consisting of two 50 minute "lessons" of lab work. Each lesson set has one pre-lab reading assignment and one writing assignment for its corresponding two lessons. One and a half to 2-hour lab classes should be able to complete one set, while 50-minute labs would complete one lesson (1/2 lesson set) per session. It is of course impossible to set a time frame for each student in a given lab. It is natural that some students will finish early, while others will need more than the suggested time frame. Chapters 6 and 13 have two lesson sets each, while Chapters 8 and 10 have only one 50-minute lesson each. Your instructor may choose individual labs based on the needs of the class. For instructors teaching semester labs that meet for 50 minutes once a week but still want to complete one set (one chapter of the text in most cases) each week, a suggested listing of labs is given at the end of this introductory section. A suggested listing of labs is also given for semester 2-hour labs. Lessons and labs correspond to the chapters in the text.

Since student generated code should be attempted only after exposure to a complete concept (chapter or set), these assignments are given at the end of the second lesson of the lesson set. Most lesson sets contain three such assignments so that the instructor may select different assignments for different sections or semesters (quarters). A few chapters have one somewhat sophisticated student generated code assignment. Other programming assignments can be found at the end of each chapter in the text. The labs are self-contained with solutions to each problem, thus enabling an instructor to use the material as an independent self-study course as well as a formalized lab. Your instructor has electronic solutions to all the programs in the labs.

Each lesson set consists of the following:

Pre-lab Reading Assignment. This will prepare the students for the material presented in the lab. This section gives a good review and additional examples, some of which are used in the following labs. Students should thus be required to read this section before coming to lab.

Pre-lab Writing Assignment. These consist of short and easy questions on the reading material so that the instructor can make sure the student did the pre-lab reading.

Two Lessons of Lab Assignments. These are done during the lab time, one lesson per hour.

Possible solutions to all lab assignments are provided on the instructor's disk.

The lessons sometimes refer to non-returning functions as procedures. Although this is not in keeping with C++ terminology, it is often pedagogically better to make this distinction both for exposure to the general terminology found in the field and to describe the unique purpose of a value returning function.

To the Instructor. . .

The teacher's notes consist of :

Objectives for Students. The objectives listed in the teacher's guide are similar to the Purpose sections given in the student manual. The objectives listed are geared more for the lab work while the student Purpose section covers the reading material as well. In some sets they are the same. The objectives are listed for the convenience of the instructor.

Assumptions. This section gives a brief list of what students should already know before attempting this lab. It is generally assumed that the students have completed and understood the previous lessons and that they have read and understood the pre-lab reading material for the current lesson set.

Pre-lab Writing Assignment Solutions. This section contains the answers to the Pre-lab Writing Assignment.

Lab Assignments. This section first lists the labs and then gives a more detailed description of each lab. Labs are broken into the lessons in which they are assigned.

Instructors are encouraged to pick and choose labs based on the needs of the individual classes. The following is a suggested order of 15 lab[1] assignments (one semester) for instructors teaching 50-minute labs that meet just once a week, but would still like to cover one chapter for most of the weeks. In essence, 50-minute labs should just do section A of each lesson set. Student generated code assignments could also be given as homework assignments.

Week 1	Lesson Set 1	Lab 1.1	Lab 1.2	Lab 1.3	Lab 1.4 (Optional Homework)
Week 2	Lesson Set 2	Lab 2.1	Lab 2.2		Lab 2.3 (Optional Homework)
Week 3	Lesson Set 3	Lab 3.1	Lab 3.2	Lab 3.3	Lab 3.4 (Optional Homework)
Week 4	Lesson Set 4	Lab 4.1	Lab 4.2	Lab 4.3	Lab 4.4 (Optional Homework)
Week 5	Lesson Set 5	Lab 5.1	Lab 5.2		Lab 5.3 (Optional Homework)
Week 6	Lesson Set 6 Part 1	Lab 6.1	Lab 6.2		Lab 6.3 (Optional Homework)
Week 7	Lesson Set 6 Part 2	Lab 6.5	Lab 6.6		Lab 6.7 (Optional Homework)
Week 8	Lesson Set 7	Lab 7.1	Lab 7.2		Lab 7.3 (Optional Homework)
Week 9	Lesson 8	All Lesson 8 labs			
Week 10	Lesson Set 9	Lab 9.1	Lab 9.2		Lab 9.3 (Optional Homework)

continues

[1] For 14 weeks, just ignore Week 15.

Week 11	Lesson 10	All Lesson 10 labs		
Week 12	Lesson Set 11	Lab 11.1	Lab 11.2	Lab 11.3 (Optional Homework)
Week 13	Lesson Set 12	Lab 12.1	Lab 12.2	Lab 12.3 (Optional Homework)
Week 14	Lesson Set 13 Part 1	Lab 13.1	Lab 13.2	Lab 13.3 (Optional Homework)
Week 15	Lesson Set 13 Part 2	Lab 13.5	Lab 13.6	Lab 13.7 (Optional Homework)

The following is a schedule for semester 2-hour weekly labs

Week 1	Lesson Set 1
Week 2	Lesson Set 2
Week 3	Lesson Set 3
Week 4	Lesson Set 4
Week 5	Lesson Set 5
Week 6	Lesson Set 6.1
Week 7	Lesson Set 6.2
Week 8	Lesson Set 7
Week 9	Lesson Set 9
Week 10	Lesson 8 & Lesson 10
Week 11	Lesson Set 11
Week 12	Lesson Set 12
Week 13	Lesson Set 13.1
Week 14	Lesson Set 13.2

LESSON SET

1 Introduction to Programming and the Translation Process

PURPOSE

1. To become familiar with the login process and the C++ environment used in the lab
2. To understand the basics of program design and algorithm development
3. To learn, recognize and correct the three types of computer errors:

   syntax errors
   run time errors
   logic errors
4. To learn the basics of an editor and compiler and be able to compile and run existing programs
5. To enter code and run a simple program from scratch

PROCEDURE

1. Students should read the Pre-lab Reading Assignment before coming to lab.
2. Students should complete the Pre-lab Writing Assignment before coming to lab.
3. In the lab, students should complete Labs 1.1 through 1.4 in sequence. Your instructor will give further instructions as to grading and completion of the lab.

Contents	Pre-requisites	Approximate completion time	Page number	Check when done
Pre-lab Reading Assignment		20 min.	2	
Pre-lab Writing Assignment	Pre-lab reading	10 min.	6	
Lesson 1A				
Lab 1.1 Opening, Compiling and Running Your First Program	Pre-lab reading	20 min. (Including overview of local system)	7	
Lab 1.2 Compiling a Program with a Syntax Error	Familiarity with the environment Finished Lab 1.1	15 min.	7	

continues

Lab 1.3 Running a Program with a Run Time Error	Understanding of the three types of errors	15 min.	8
Lesson 1B			
Lab 1.4 Entering a New Program and Working with Logic Errors	Understanding of errors	15 min.	9
Lab 1.5 Writing Your First Program	Finished Labs 1.1 through 1.4	30 min.	10

PRE-LAB READING ASSIGNMENT

Computer Systems

A **computer system** consists of all the components (hardware and software) used to execute the desires of the computer user. **Hardware** is the electronic physical components that can retrieve, process and store data. It is generally broken down into five basic components:

Central Processing Unit (C.P.U.) This is the unit where programs are executed. It consists of the **control unit**, which oversees the overall operation of program execution and the **A.L.U.** (Arithmetic/Logic Unit), which performs the mathematical and comparison operations.

Main Memory — where programs and data are stored for use by the CPU

Secondary Storage — where programs and data are filed (stored) for use at a later time

Input Devices — devices used to get programs and data into the computer (keyboard)

Output Devices — devices used to get programs and data from the computer (printer)

Software consists of a sequence of instructions given to perform some pre-defined task. These labs concentrate on the software portion of a computer system.

Introduction to Programming

A **computer program** is a series of instructions written in some computer language that performs a particular task. Many times beginning students concentrate solely on the language code; however, quality software is accomplished only after careful design that identifies the needs, data, process and anticipated outcome. For this reason it is critical that students learn good design techniques before attempting to produce a quality program. Design is guided by an **algorithm**, which is a plan of attacking some problem. An algorithm is used for many aspects of tasks, whether a recipe for a cake, a guide to study for an exam or the specifications of a rocket engine.

Problem example: Develop an algorithm to find the average of five test grades.

An algorithm can begin with a general broad statement of the problem.

Find the average of Five Test Grades

From here we can further refine the statement by listing commands that will accomplish our goal.

Read in the Grades	Find the Average	Write out Average

Each box (called a node) may or may not be refined further depending on its clarity to the user. For example: **Find the Average** may be as refined as an experienced programmer needs to accomplish the task of finding an average; however, students learning how to compute averages for the first time may need more refinement about how to accomplish the goal. This refinement process continues until we have a listing of steps understandable to the user to accomplish the task. For example, **Find the Average** may be refined into the following two nodes.

Total= sum of 5 grades	Average=Total/5

Starting from left to right, a node that has no refinement becomes part of the algorithm. The actual algorithm (steps in solving the above program) is listed in bold.

Find the Average of Five Test Grades

Read in the Grades

Find the Average

Total = sum of 5 grades
Average = Total / 5

Write Out the Average

From this algorithm, a program can be written in C++.

Translation Process

Computers are strange in that they only understand a sequence of 1s and 0s. The following looks like nonsense to us but, in fact, is how the computer reads and executes everything that it does:

10010001111010101110010001110001000

Since computers only use two numbers (1 and 0), this is called **binary** code. One can imagine how complicated programming would be if we had to learn this very complex language. That in fact was how programming was done many years ago; however, today we are fortunate to have what are called **high level languages** such as C++. These languages are geared more for human understanding and thus make the task of programming much easier. However, since the computer only understands low level binary code (often called machine code), there must be a translation process to convert these high level languages to machine code. This is often done by a **compiler**, which is a software package that translates high level languages into machine code. Without it we could not run our programs. The figure below illustrates the role of the compiler.

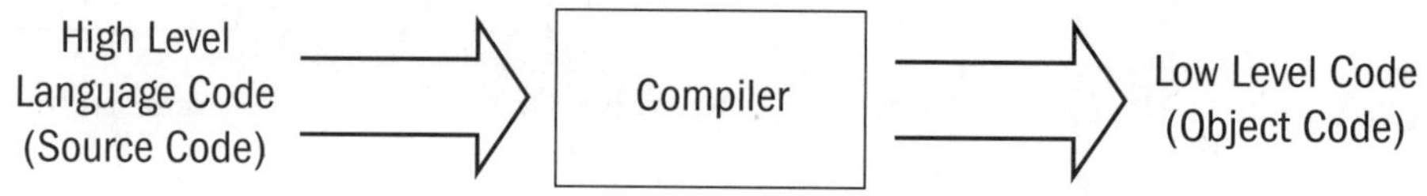

The compiler translates source code into object code. The type of code is often reflected in the extension name of the file where it is located.

Example: We will write source (high level language) code in C++ and all our file names will end with `.cpp`, such as:

`firstprogram.cpp`   `secondprogram.cpp`   etc.

When those programs are compiled, a new file (object file) will be created that ends with `.obj`, such as:

`firstprogram.obj`   `secondprogram.obj`   etc.

The compiler also catches grammatical errors (called syntax errors) in the source code. Just like English, all computer languages have their own set of grammar rules that have to be obeyed. If we turned in a paper with a proper name (like John) not capitalized, we would be called to task by our teacher, and probably made to correct the mistake. The compiler does the same thing. If we have something that violates the grammatical rules of the language, the compiler will give us error messages. These have to be corrected and a grammar error free program must be submitted to the compiler before it translates the source code into machine language. In C++, for example, instructions end with a semicolon. The following would indicate a syntax error:

```
cout << "Hi there" << endl
```

Since there is no semicolon at the end, the compiler would indicate an error, which must be corrected as follows:

```
cout << "Hi there" << endl;
```

After the compile process is completed, the computer must do one more thing before we have a copy of the machine code that is ready to be executed. Most programs are not entirely complete in and of themselves. They need other programs previously written that perform certain routine operations such as data input and output. Our programs need these attachments in order to run. This is the function of the **linking process**. Suppose you are writing a term paper on whales and would like a few library articles attached to your report. You would go to the library, get a copy of the articles (assuming it would be legal to do so), and attach them to your paper before turning it in. The **linker** does this to your program. It goes to a "software library" of programs and attaches the appropriate code to your program. This produces what is called the executable code, generated in a file that often ends with `.exe`.

Example: `firstprogram.exe`   `secondprogram.exe`

The following figure summarizes the translation process:

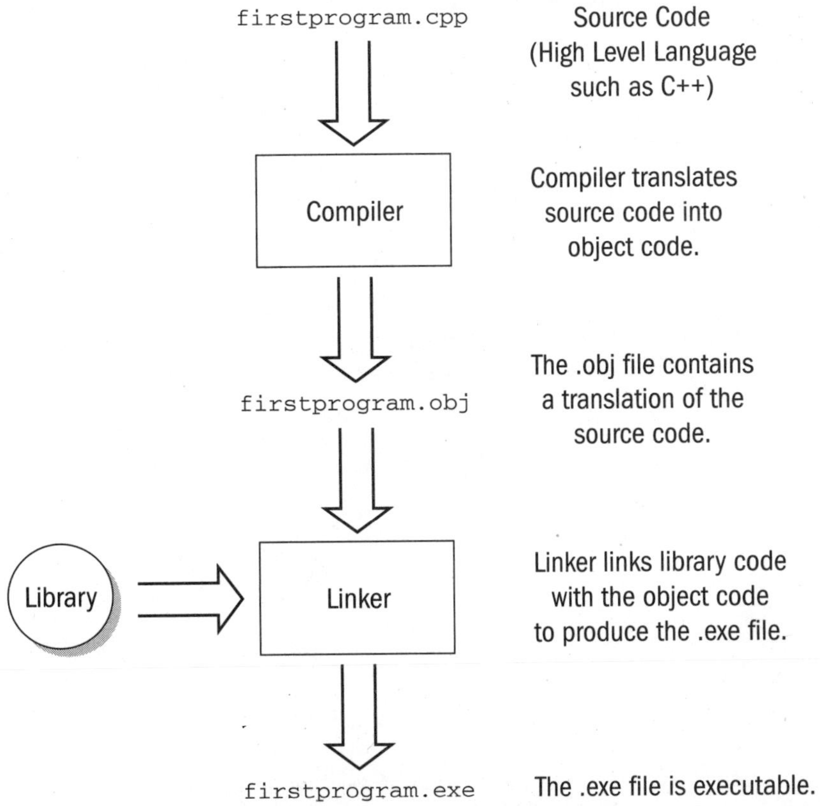

Once we have the executable code, the program is ready to be run. Hopefully it will run correctly and everything will be fine; however that is not always the case. During "run time", we may encounter a second kind of error called a **run time error**. This error occurs when we ask the computer to do something it cannot do. Look at the following sentence:

You are required to swim from Naples, Italy to New York in five minutes.

Although this statement is grammatically correct, it is asking someone to do the impossible. Just as we cannot break the laws of nature, the computer cannot violate the laws of mathematics and other binding restrictions. Asking the computer to divide by 0 would be an example of a run time error. We would get executable code; however, when the program tries to execute the command to divide by 0, the program will stop with a run time error. Run time errors, particularly in C++, are usually more challenging to find than syntax errors.

Once we run our program and get neither syntax nor run time errors, are we free to rejoice? Not exactly. Unfortunately, it is now that we may encounter the worst type of error: the dreaded **Logic error**. Whenever we ask the computer to do something, but mean for it to do something else, we have a logic error. Just as there needs to be a "meeting of the minds" between two people for meaningful communication to take place, there must be precise and clear instructions that generate our intentions to the computer. The computer only does what we ask it to do. It does not read our minds or our intentions! If we ask a group of people to cut down the tree when we really meant for them to trim the bush, we have a communication problem. They will do what we ask, but what we asked and what we wanted are two different things. The same is true for the computer. Asking it to multiply by 3 when we want something doubled is an example of a logic error. Logic errors are the most difficult to find and correct because there are no error messages to help us locate the problem. A great deal of programming time is spent on solving logic errors.

Integrated Environments

An integrated environment is a software package that bundles an editor (used to write programs), a compiler (that translates programs) and a run time component into one system. For example, the figure below shows a screen from the Microsoft Visual C++ integrated environment.

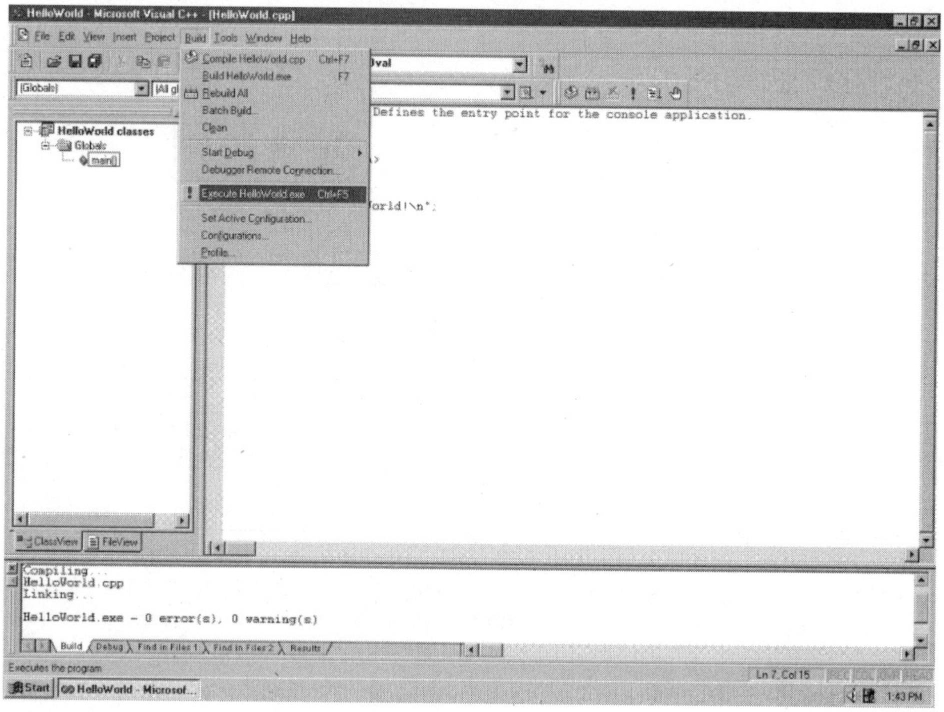

Other systems may have these components separate which makes the process of running a program a little more difficult. You should also be aware of which Operating System you are using. An **Operating System** is the most important software on your computer. It is the "grand master" of programs that interfaces the computer with your requests. Your instructor will explain your particular system and C++ environment so that you will be able to develop, compile and run C++ programs on it.

PRE-LAB WRITING ASSIGNMENT

Fill-in-the-Blank Questions

1. Compilers detect ________________ errors.
2. Usually the most difficult errors to correct are the ________________ errors, since they are not detected in the compilation process.
3. Attaching other pre-written programs to your program is done by the ________________ process.
4. ________________ code is the machine code consisting of ones and zeroes that is read by the computer.
5. Dividing by zero is an example of a ________________ error.

Learn the Environment That You Are Working In

The following information may be obtained from your instructor.

1. What operating system are you using?
2. What C++ environment are you working in?
3. If you are not working in an integrated environment, what are the compile, run and edit commands that you will need?

LESSON 1A

Your instructor may assign either appendix A or appendix B depending on your environment. Appendix A is for labs using Visual C++ and Appendix B is for labs using UNIX. If you are using an environment other than these two, your instructor will give you instructions for this first lesson and ask you to complete Lab 1.1 below.

LAB 1.1 Opening, Compiling and Running Your First Program

Exercise 1: Logon to your system based on your professor's instructions.

Exercise 2: Bring in the `firstprog.cpp` program.

Exercise 3: Compile the program.

Exercise 4: Run the program and write what is printed on the screen.

The code of `firstprog.cpp` is as follows:

```
// This is the first program that just writes out a simple message

// PLACE YOUR NAME HERE

#include <iostream.h>

void main ()

{

    cout << "Now is the time for all good men" << endl;
    cout << "To come to the aid of their party" << endl;

}
```

LAB 1.2 Compiling a Program with a Syntax Error

Exercise 1: Bring in program `semiprob.cpp` from the Lab 1 folder.

Exercise 2: Compile the program. Here we have our first example of the many syntax errors that you no doubt will encounter in this course. The error message you receive may be different depending on the system you are using, but the compiler insists that a semicolon is missing somewhere. Unfortunately, where the message indicates that the problem exists, and where the problem actually occurs may be two different places. To correct

the problem place a semicolon after the line `cout << "Today is a great day for Lab"`.
Most syntax errors are not as easy to spot and correct as this one.

Exercise 3: Re-compile the program and when you have no syntax errors, run the program and input 9 when asked. Record the output.

Exercise 4: Try running it with different numbers. Record your output.

Do you feel you are getting valid output?

The code of `semiprob.cpp` is as follows:

```
// This program demonstrates a compile error

// PLACE YOUR NAME HERE

#include <iostream.h>

void main()

{
   int number;
   float total;


   cout << "Today is a great day for Lab"
   cout << endl << "Let's start off by typing a number of your choice" << endl;
   cin >> number;

   total = number * 2;
   cout << total << " is twice the number you typed" << endl;

}
```

LAB 1.3 Running a Program with a Run Time Error

Exercise 1: Bring in program `runprob.cpp` from the Lab 1 folder.

Exercise 2: Compile the program. You should get no syntax errors.

Exercise 3: Run the program. You should now see the first of several run time errors. There was no syntax or grammatical error in the program; however, just like commanding someone to break a law of nature, the program is asking the computer to break a law of math by dividing by zero. It cannot be done. On some installations, you may see this as output that looks very strange. Correct this program by having the code divide by 2 instead of 0.

Exercise 4: Re-compile and run the program. Type 9 when asked for input. Record what is printed.

Exercise 5: Run the program using different values. Record the output.

Do you feel that you are getting valid output?

The code of `runprob.cpp` is as follows:

```
// This program will take a number and divide it by 2

// Place your name here

#include <iostream.h>

void main()

{

    float number;
    int divider;

    divider = 0;

    cout << "Hi there" << endl;
    cout << "Please input a number and then hit return" << endl;
    cin >> number;

    number = number / divider;

    cout << "Half of your number is " << number << endl;

}
```

LESSON 1B

LAB 1.4 Entering a New Program and Working with Logic Errors

Exercise 1: Key in the following code. You are not expected to understand how it works; just copy it as shown.

```
// This program takes two values from the user and then swaps them before printing the
// the values. The user will be prompted to enter both numbers

//  Place your name here

#include <iostream.h>

void main()

{

    float firstnumber;
    float secondnumber;

    //prompt user to enter the first number
```

continues

```
    cout << "Enter the first number" << endl;
    cout << "Then hit enter" << endl;
    cin >> firstnumber;

    //promt user to enter the second number

    cout << "Enter the second number" << endl;
    cout << "Then hit enter" << endl;
    cin >> secondnumber;

    //Echo print the input

    cout << endl << "You inputed the numbers as " << firstnumber << " and "
         << secondnumber << endl;

     //Now we will swap the values

    firstnumber = secondnumber;
     secondnumber = firstnumber;

    //Output the values

    cout << "After swapping, the first number has the value of " << firstnumber
         << " which was the value of the second number." << endl
         << "The second number has the value of " << secondnumber
         << " which was the value of the first number." << endl;

}
```

Exercise 2: Save this program as `logicprob.cpp`.

Exercise 3: Compile this program. If you have compile errors, go back and check your work. Make sure everything you typed is exactly the way it is in the book.

Exercise 4: When you have no syntax errors. Run the program. What is printed?

Exercise 5: This program has no syntax or run time errors, but it certainly has a logic error. This logic error may not be easy to find. Most logic errors create a challenge for the programmer. Your instructor may ask you not to worry about finding and correcting the problem at this time.

LAB 1.5 Writing Your First Program (Optional)

Exercise 1: Develop a design that leads to an algorithm and a program that will read in a number that represents the number of kilometers traveled. The output will convert this number to miles. 1 kilometer = 0.621 miles. Call this program `kilotomiles.cpp`.

Exercise 2: Compile the program. If you get compile errors, try to fix them and re-compile until you are syntax error free.

Exercise 3: Run the program. Is your output what you expect from the input you gave? If not, try to find and correct the logic error and run the program again. Continue this process until you have a program that produces the correct result.

LESSON SET

2 Introduction to the C++ Programming Language

PURPOSE

1. To briefly introduce the C++ programming language
2. Show the use of memory in programming
3. Discuss the difference between variable and constant declarations
4. Introduce various data types:
    a. Integer
    b. Character
    c. Floating point
    d. Boolean
5. To introduce the assignment statement
6. To demonstrate the use of arithmetic operators

PROCEDURE

1. Students should read the Pre-lab Reading Assignment before coming to lab.
2. Students should complete the Pre-lab Writing Assignment before coming to lab.
3. In the lab, students should complete Labs 2.1 through 2.4 in sequence. Your instructor will give further instructions as to grading and completion of the lab.

Contents	Pre-requisites	Approximate completion time	Page number	Check when done
Pre-lab Reading Assignment		20 min.	12	
Pre-lab Writing Assignment	Pre-lab reading	10 min.	17	
Lesson 2A				
Lab 2.1 Working with the `cout` Statement	Pre-lab reading	20 min.	17	
Lab 2.2 Working with Constants, Variables and Arithmetic Operators	Understanding of variables and operators	30 min.	18	

continues

Lesson 2B				
Lab 2.3 Rectangle Area and Perimeter	Understanding of basic components of a program	45 min.	19	

PRE-LAB READING ASSIGNMENT

The C++ Programming Language

Computer programming courses generally concentrate on program design that can be applied to any number of programming languages on the market. It is imperative, however, to apply that design to a particular language. This course uses C++, a popular object-oriented language, for that purpose.

For now, we can think of a C++ program as consisting of two general divisions: header and main. The header, or global section, gives preliminary instructions to the compiler. It consists of comments that describe the purpose of the program, as well as information on where to get programs that will be linked (see Lesson Set 1).

```
// This program prints to the screen the words: PI = 3.14
//                                               Radius = 4
//                                               Circumference = 25.12

#include <iostream.h>

const double PI = 3.14;

void main()
{
      float radius;
      radius = 4.0;

      cout << "PI = " << PI << endl;
      cout << "Radius = " << radius << endl;
      cout << "Circumference = " << 2 * PI * radius << endl;

}
```

Everything in bold (everything above the `void main()` statement) is considered the header or global section. Everything in italics (everything below and including the `void main()` statement) is the main section.

Comments are included in every program to document what a program does and how it operates. These statements are ignored by the computer but are most valuable to the programmers who must update or fix the program. In C++, comments begin with `//` which is an indication to the compiler to ignore everything from the `//` to the end of the line. Comments can also cross line boundaries by beginning with /* and ending with */. Notice that the first three lines of the above program all begin with // and thus are comments. Those same lines could also have been written as the following:

```
/*   This program prints to the screen the words: PI = 3.14
                                                  Radius = 4
                                                  Circumference = 25.12

*/
```

The next statement, the #include statement, is a command that directs the compiler to link the appropriate library to the program.

```
#include <iostream.h>
```

Recall from Lesson Set 1, that every program needs other programs attached to it so that it may execute properly. Your instructor will generally tell you what libraries are needed for each particular programming assignment; however, in time you will learn this task for yourself.

Every C++ program has a **main** function which indicates the start of the executable instructions. Every main must begin with a left brace { and end with a right brace }. The statements inside those braces will be explained as we progress through this lesson.

Memory

Memory storage is the collection of locations where instructions and data that are used by the instructions are temporarily stored. Recall from Lesson Set 1 that a computer only understands a sequence of 1s and 0s. These are binary digits or bits (**BInary digiTs**). Eight of these brought together are called a **byte**, which is the most common unit of storage. These chunks of memory can be thought of as hotel mailboxes at the registration desk. The size of each of those boxes indicates the type of mail that can be stored there. A very small mailbox may only hold notes or postcards. Larger mailboxes can hold letters, while even larger ones can hold packages. Each mailbox is identified by a number or name of an occupant. We have identified two very important attributes of these mailboxes: the name or number, which indicates the mailbox that is being referenced, and the size, which indicates what type of "data" can be placed there.

Example: **postcards Jim** could be an indication that the mailbox called Jim can only hold postcards, while the statement **packages Mary** would indicate that the mailbox called Mary could hold large packages. Memory locations in a computer are identified by the same two attributes: data type and name.

Much of programming is getting data to and from memory locations and thus it is imperative that the programmer tell the computer the name and data type of each memory location that he or she intends to use. In the sample program the statement **float radius** does just that. float is a data type that indicates what kind of data can be stored and radius is the name for that particular memory location.

Variables and Constants

The ability to change or not change the data stored could be a third attribute of these memory locations. Components of memory in which data values stored can change during the execution of the program are called **variables**. These should never be declared in the header or global section of the program. In our example, `radius` (which is declared as having a `float` data type) is declared in the main section. Components of memory in which data values stored are initialized once and never changed during the execution of the program are called **constants**. They are often declared in the global section and are preceded (in C++) by the word **`const`**.

Identifiers in C++

Identifiers are used to name variables, constants and many other components of a program. They consist exclusively of letters, digits and the underscore _ character. They cannot begin with a digit and cannot duplicate reserved words used in C++ such as **`int`** or **`if`**. All characters in C++ are case sensitive, thus memory locations called `simple`, `Simple`, and `SIMPLE` are three distinct locations. It has become standard practice among programmers to make constants all uppercase and variables predominantly lowercase characters.

The statement **`const double PI = 3.14;`** in our sample program is contained in the global section. It defines a memory location called `PI` to be a constant holding a double (a data type discussed shortly) value equal to 3.14 which will NOT change during the execution of the program.

The statement **`float radius;`** in the sample program is contained in the main section. It defines a variable memory location called `radius` that will hold a floating point data type (type discussed shortly) which can be changed during the execution of the program.

Both of these statements are called **declarations** because they reserve by name enough memory to hold the data type described.

Variables, like constants, can be given an initial value at declaration, but that value is not permanent and can be altered. For example:

```
int count = 7;      //Defines a variable memory location called count that
                    //has initially the value of 7
count = count + 1;  //count is now altered
```

Data Types

As noted earlier, computer memory is composed of units identified by a name (like the room number of a hotel mailbox) and a data type. This data type indicates what kind of data can be stored, thus setting the size of that location.

Integer Data Type

Integers are real numbers that do not contain any fractional component. They take up less memory than numbers with fractional components. C++ has three data types that are integers: **`short`**, **`int`** and **`long`**. The difference is strictly in the amount of memory (bytes) they reserve: short reserving the least and long reserving the most. Larger integers may need the long data type.

The following three statements declare integer variables in C++:

```
short count;
int sum;
long total;
```

Floating Point Data Type

In computer science 3 = 3.0 is not a true statement. The number on the left is an integer and the number on the right is a real, or floating point, number (a number that has a fractional component). Although mathematically the two are equal, the computer stores them as different data types. C++ uses both **`float`** and **`double`** to indicate floating point numbers, with double using more memory than float.

The following two statements declare floating point variables in C++.

```
float average;
double nationaldebt;
```

Character Data Type

Character data includes the letters of the alphabet (upper and lower cases), the digits 0–9 and special characters such as ! ? . , *. All these symbols combined are called alphanumeric. Each character data is enclosed with single quotes to distinguish it from other data types. Thus `'8'` is different than `8`. The first is a character while the second is an integer.

Boolean Data Type

The Boolean data type, named after the mathematician George Boole, allows only two values: true or false, which are internally represented as one and zero, respectively. The statement **`bool found;`** declares a boolean data type in C++ that can have either the value true or the value false.

Fundamental Instructions

Most programming languages, including C++, consist of five fundamental instructions from which all code can be generated.

1. **Assignment Statements**: These statements place values in memory locations. The left side of an assignment statement consists of one and only one variable. The right side consists of an **expression**. An expression can be any manipulation of **literal** numbers (actual numbers such as 7 or 38, etc.), or the contents of constants and/or variables, that will "boil down" to one value. That value is placed in the memory location of the variable on the left. C++ uses = as the separator between the left and right side of the assignment statement. Those new to programming often get this confused with equality; however = in C++ is not equality but rather the symbol to indicate assignment. The = in C++ is read as "is assigned the value of".

Example:

```
int count;

int total;

total = 10;        // 10 is a literal that is placed in the memory
                   // location called total

count = 3 + 4;     // The right hand side of the statement is evaluated to
                   // produce a 7. count is assigned the value of 7.

total = total + count; // The right hand side is evaluated (10 +7)
                       // and 17 is placed in the memory location called
                       // total.
```

This last statement may seem a bit confusing. Starting with the right side, it says to get the value that is in `total` (10 in this case) and add it to the value that is in `count` (7 in this case), then store that combined sum (17) in the memory location called `total`. Notice that `total`, which was initially 10, gets changed to 17.

2. **Output Statements**: These instructions send information from the computer to the outside world. This information may be sent to the screen or to some file. In C++ the `cout <<` statement sends information to the screen. The `#include <iostream.h>` must be in the header for `cout` to be used.

```
cout << total;
```

The above statement sends whatever value is stored in the variable `total` to the screen.

We can output literal strings (such as "Hello") by inclosing them in double quotes.

The << acts as a separator for multiple outputs.

```
cout << "The value of total is " << total << endl;
```

The `endl` statement causes the cursor to be moved to the beginning of the next line.

The remaining three fundamental instructions will be explained in future labs.

3. **Input Statements**: These statements bring in data to the computer. (Lesson Set 3)

4. **Conditional Statements**: These instructions test conditions to determine which path of instructions to execute. (Lesson Set 4)

5. **Loops**: These instructions indicate a repetition of a series of instructions. (Lesson Set 5)

C++ uses the semicolon as a statement terminator.

Arithmetic Operators

Programming has the traditional arithmetic operators:

addition +
subtraction -
multiplication *
division /
modulus %

Integer division occurs when an integer (or number stored in a variable declared to be integer) is divided by an integer (or number stored in a variable declared to be integer). The result is always an integer. Thus 9/2 will give the answer 4 not 4.5! For this reason there are two division operations for integer numbers. The % is used as an operator for integer division to give the remainder of a division operation. 9/2 gives 4 while 9 % 2 gives 1 (the remainder of the division).

Example:

```
int count = 9;
int div = 2;
int remainder;
int quotient;

quotient = count / div;
remainder = count % div;
```

The value of 9/2 is 4 with remainder 1.

Do you see that quotient gets the value of 4 and remainder gets the value of 1? Further discussion of arithmetic operators is given in Lesson Set 3.

You should go back and review the sample program on the first page of the Pre-lab Reading Assignment. By now you should understand most of the statements.

PRE-LAB WRITING ASSIGNMENT

Fill-in-the-Blank Questions

1. A _______________ is a memory location whose value cannot change during the execution of the program.
2. _______________ is a data type that only holds numbers with no fractional component.
3. _______________ is a data type that holds numbers with fractional components.
4. _______________ is an arithmetic operation that gives the remainder of a division problem.
5. `cout <<` is an example of the _______________ fundamental instruction.
6. _______________ data types only have two values: true and false.
7. One byte consists of _______________ bits.
8. // or /* in C++ indicates the start of a _______________.
9. A _______________ is a memory location whose value can change during the execution of the program.

LESSON 2A

LAB 2.1 Working with the `cout` Statement

Exercise 1: Retrieve program `name.cpp` from the Lab 2 folder.

Fill in the code so that the program will do the following:

Write your first and last name on one line.
Write your address on the next line (recall the function of the `endl` statement).
Write your city, state and zip on the next line.
Write your telephone number on the next line.
Remember that to output a literal, such as "Hello", you must use quotes.
Compile and run the program.

Example: Deano Beano
123 Markadella Lane
Fruitland, Md. 55503
489-555-5555

The code for `name.cpp` is as follows:

```
// This program will write the name, address and telephone number of the
// programmer

// PLACE YOUR NAME HERE

#include <iostream.h>

void main()
{

   // Fill in this space to write your first and last name
   // Fill in this space to write your address (on new line)
   // Fill in this space to write your city, state and zip (on new line)
   // Fill in this space to write your telephone number (on new line)

}
```

Exercise 2: Change the program so that three blank lines separate the telephone number from the address. Compile and run the program.

Exercise 3: Change the program so that the following (but with your name and address) is printed. Try to get the spacing just like the example. Compile and run the program.

```
************
    Programmer: Deano Beano
                123 Markadella Lane
                Fruitland, Md. 55503

     Telephone: 489-555-5555

************
```

LAB 2.2 Working with Constants, Variables and Arithmetic Operators

Exercise 1: Bring in the file `circlearea.cpp` from the Lab 2 folder.

The code of `circlearea.cpp` is as follows:

```
// This Program will output the circumference and area
// of the circle with a given radius

// PLACE YOUR NAME HERE

#include <iostream.h>

const  double PI = 3.14;
const  double RADIUS = 5.4;
```

```
void main()

{
    _________ areaofcircle;             //declaration of areaofcircle

    float  circumference;               //declaration of circumference

    circumference = 2 * PI * RADIUS;     // finds the circumference

    areaofcircle =_____________;         // find the area

    // Fill in the code for the cout statement that will output (with description) the
    // circumference

    // Fill in the code for the cout statement that will output (with description) the
    // area of the circle

}
```

Exercise 2: Fill in the blanks and the `cout` statements so that the output will produce the following:

The circumference of the circle is 33.912
The area of the circle is 91.5624

Exercise 3: Change the data type of circumference from `float` to `int`. Run the program and record the results.

The circumference of the circle is _______________.

The area of the circle is _______________.

Explain what happened to get the above results.

LESSON 2B

LAB 2.3 Rectangle Area and Perimeter

Exercise 1: Using Lab 2.2 as an example, develop a program that will determine the area and perimeter of a rectangle. The length and width can be given as constants. (`LENGTH=8 WIDTH=3`)

Exercise 2: Compile and run your program. Continue to work on it until you get the following output.

The area of the rectangle is 24
The perimeter of the rectangle is 22

LESSON SET

3 Expressions, Input, Output and Data Type Conversions

PURPOSE

1. To learn input and formatted output statements
2. To learn data type conversions (coercion and casting)
3. To work with constants and mathematical functions
4. To briefly introduce the concept of files

PROCEDURE

1. Students should read the Pre-lab Reading Assignment before coming to lab.
2. Students should complete the Pre-lab Writing Assignment before coming to lab.

Contents	Pre-requisites	Approximate completion time	Page number	Check when done
Pre-lab Reading Assignment		20 min.	22	
Pre-lab Writing Assignment	Pre-lab reading	10 min.	27	
LESSON 3A				
Lab 3.1 Working with the `cin` Statement	Confidence in use of data types	15 min.	27	
Lab 3.2 Formatting Output	Basic understanding of `cout` and formatted output	15 min.	29	
Lab 3.3 Arithmetic Operations and Math Functions	Understanding of pre-defined functions `pow` and `sqrt`	20 min.	30	
LESSON 3B				
Lab 3.4 Working with Type Casting	Understanding of type casting (implicit and explicit data conversion)	20 min.	31	

continues

Lab 3.5 Reading and Writing to a File	Basic understanding of reading and writing files	15 min.	32
Lab 3.6 Student Generated Code Assignments	Understanding of all concepts (except files) covered in this section.	30 min.	33

PRE-LAB READING ASSIGNMENT

Review of the `cout` Statement

The **`cout`** statement invokes an output stream, which is a sequence of characters to be displayed to the screen.

Example: `cout << "Hi there";`

The **insertion operator** << inserts the string of characters Hi there into the output stream that goes to the screen. The `cout` statement can be thought of as an **ofstream** (output stream) data type.

Input Instructions

Just as the `cout` statement transfers information from the computer to the "outside" world, the `cin` statement transfers data into the computer from the keyboard.

Example: `cin >> grade;`

The **extraction operator** >> extracts an item from the input stream. In this case, since `grade` is an integer, this instruction will wait for an integer to be entered at the keyboard and then will place that number in the memory location called `grade`.

Just as `cout` is of type ofstream, `cin` is considered to be an **ifstream** (input stream) data type. In order to use `cin` and `cout` in a C++ program, the `#include <iostream.h>` directive should be included in the header. The >> extraction operator also serves as a separator between input variables, allowing more than one memory location to be loaded with a single `cin` instruction. The values read must be the same data type as their corresponding variables, although a floating point variable could receive an integer since it will make the conversion. Conversion is discussed later in this lesson.

Example:

```
float rate;
float hours;

cin >> rate >> hours;
```

The `cin` statement will wait for two floating point numbers (separated by at least one blank space) to be input from the keyboard. The first will be stored in `rate` and the second in `hours`. There is one problem with the example above; it does not indicate to the user for what data the `cin` statement is waiting. Remember that the `cin` statement is expecting data from the user at the keyboard. For this reason, every `cin` statement should be preceded by a `cout` statement that indicates to the user the data to be input. That `cout` statement is called a prompt.

Example:

```
float rate,hours;
                                //More than one variable can be declared
                                //in a statement. Multiple variables are
                                //separated by a comma.
float grosspay;

cout << "Please input the pay rate per hour"
     << " and then the number of hours worked" << endl;
cin >> rate >> hours;

grosspay = rate * hours;         //finds the grosspay

cout << endl << "The rate is = " << rate << endl;
cout << "The number of hours = " << hours << endl;
cout << "The gross pay =  " << grosspay << endl;
```

When `cin` is reading numeric data, whitespace (blank spaces or unseen control characters) are ignored and the read continues until a non-numeric character is encountered.

Formatted Output

C++ has special instructions that allow the user to control output attributes such as spacing, data format, decimal point precision as well as other features. For these special instructions to work, the `#include <iomanip.h>` instruction must be included in the header (global section).

Example:

```
cout << setprecision(2);          // This instruction will force all
                                  // floating point numbers to be rounded to
                                  // two decimal places.

cout << setiosflags(ios::fixed); // This forces the output to be in decimal
                                  // format instead of scientific notation.

cout << setiosflags(ios::showpoint); // This makes all floating point output to
                                     // show a decimal point even if the values
                                     // are whole numbers.
```

The above could all be combined into one statement:

```
cout << setiosflags(ios::fixed) << setiosflags(ios::showpoint) << setprecision(2);
```

Spacing is handled by an indication of the width of the field that the number or character is to be placed in. It can be done with the `cout.width(n);` where `n` is the width size. However it is more commonly done by the `setw(n)` within a `cout` statement.

Example:
```
float price = 9.5;
float rate = 8.76;
cout << setw(10) << price << setw(7) << rate;
```

The above statements will print the following:

```
       9.5   8.76
```

There are seven blank spaces before 9.5 and three blank spaces between the numbers. The numbers are right justified. The computer memory stores this as follows:

							9	.	5				8	.	7	6

Note: So far we have used `endl` for a new line of output. The `'\n'` is an escape sequence used as a character in a literal that will accomplish the same thing.

Example: Both of the following will do the same thing.

```
cout << "Hi there\n";          cout << "Hi there" << endl;
```

Expressions

Recall from Lesson Set 2 that the assignment statement consists of two parts: a variable on the left and an expression on the right. The expression is converted to one value that is placed in the memory location corresponding to the variable on the left. These expressions can consist of variables, constants and literals combined with various mathematical operations. It is important to remember the mathematical precedence rules which are applied when solving these expressions.

Precedence Rules in Order of Priority

1. Anything grouped in parentheses is top priority
2. Unary negation example: –8
3. Multiplication, Division and Modulus * / %
4. Addition and Subtraction + –

Example:
```
( 8 * 4/2 + 9 - 4/2 + 6 * (4+3) )
( 8 * 4/2 + 9 - 4/2 + 6 * 7 )
( 32 /2   + 9 - 4/2 + 6 * 7 )
(   16    + 9 - 4/2 + 6 * 7 )
(   16    + 9 -  2  + 6 * 7 )
(   16    + 9 -  2  + 42 )
(      25     -  2  + 42 )
(           23      + 42 )  =   65
```

Converting Algebraic Expressions to C++ Expressions

One of the challenges of learning a new computer language is the task of changing algebraic expressions to their equivalent computer instructions.

Example: `4y(3-2)y+7`

How would this algebraic expression be implemented in C++?

```
4 * y * (3-2) * y + 7
```

Other expressions are a bit more challenging. Consider the quadratic formula:

$$\frac{-b \pm \sqrt{b^2 - 4ac}}{2a}$$

We need to know how C++ handles functions such as the square root and squaring functions.

There are several predefined math library routines that are contained in the `math` library. In order to use these we must have the `#include <math.h>` directive in the header.

Exponents are handled by the `pow(number,exp)` function where `number` indicates the base and `exp` is the exponent.

2^3 `pow(2,3)`

5^9 `pow(5,9)`

Square roots are handled by `sqrt(n)`

$\sqrt{9}$ `sqrt(9)`

Look at the following C++ statement and try to determine what it is doing.

```
formula1 = ( -b + sqrt(pow(b,2) -(4 * a * c))) / (2 * a);
formula2 = ( -b - sqrt(pow(b,2) -(4 * a * c))) / (2 * a);
```

(These are the roots from the quadratic formula in C++ format.)

Data Type Conversions

Recall the discussion of data types from Lesson Set 2. Whenever an integer and a floating point variable (object) or constant are mixed in an operation, the integer is changed temporarily to its equivalent floating point. This automatic conversion is called **implicit type coercion**.

Consider the following:

```
int count;
count = 7.8;
```

We are trying to put a floating point number into an integer memory location. This is like trying to stuff a package into a mailbox that is only large enough to contain letters. Something has to give. In C++ the floating point is **truncated** (the entire fractional component is cut off) and, thus, we have loss of information.

Type conversions can be made explicit (by the programmer) by placing the value to be changed in parentheses and placing before it (to the left) the name of the new type. This is called **type casting** or **type conversion**.

Examples:

```
int count;
float sum;

count = 10.89;                    //Float to integer  This is Type coercion
                                  //10 is stored in count

count = int(10.89);               //Also float to integer; however this is
                                  //type casting
```

If two integers are divided, the result is an integer that is truncated. This can create unexpected results.

Example:
```
int num_As = 10;
int totalgrade = 50;
float percent_As;

percent_As = num_As / totalgrade;
```

In this problem we would expect `percent_As` to be .20 since 10/50 is .20. However since both `num_As` and `totalgrade` are integers, the result is integer division which gives a truncated number. In this case it is 0. Whenever a smaller integer value is divided by a larger integer value the result will always be 0. We can correct this problem by type casting.

```
percent_As = float(num_As) / totalgrade;
```

Notice that `num_As` is now converted to a float and we have a float divided by the integer `totalgrade`. That creates a type coercion into the floating number 0.20 which is now placed in `percent_As` giving us the value we expect.

Files (Optional)

Although files are discussed at length in Lesson Set 12, we give a brief introduction here for students wishing to use files early in the course.

Input can come from files instead of from the keyboard and output can go to files instead of to the screen. The `#include <fstream.h>` should be included in the heading so that files can be created and accessed. A file containing data to be input to the computer should be declared as an **ifstream** data type while an output file should be declared as **ofstream**.

Sample Program 3.1

Suppose we have a data file called `grades.dat` that contains 3 grades and want to take those grades and output them to an output file that we will call `finalgrade.out`. The following code shows how this can be done in C++:

```
#include <fstream.h>                    // allows the use of data files and
                                        // output files.

void main()
{
   float grade1, grade2, grade3;     // declares 3 float variables

   ifstream dataFile;                // declares an input stream.
                                     // dataFile is the"internal" name that we
                                     // will use for the file grades.dat

   ofstream outFile;                 // declares an output stream.  outFile is
                                     // the"internal" name that we will use
                                     // for the output file called
                                     // finalgrade.out

   outFile << setiosflags(ios::fixed) << setiosflags(ios::showpoint);
                                     // We can use these for output files as
                                     // well as for cout.

   dataFile.open("grades.dat");      // This ties the internal name, dataFile,
                                     // to the actual file, grades.dat
```

```
    outFile.open("finalgrade.out")    //This ties the internal name, outFile, to
                                      //the actual real name, finalgrade.out, of
                                      //the file.

    dataFile >> grade1 >> grade2 >> grade3;
    outFile << grade1 << endl;
    outFile << grade2 << endl;
    outFile << grade3 << endl;

}
```

PRE-LAB WRITING ASSIGNMENT

Fill-in-the-Blank Questions

1. What is the final value (in C++) of the following expression?

   `(5 - 16 / 2 * 3 + (3 + 2 / 2) - 5)` ______________

2. How would the following expression be written in C++?

   $2x + 3^4$

   ____

3. Implicit conversion is also known as data type ______________.
4. Explicit type conversion is also known as type ______________.
5. List the preprocessor directive that must be included for `cin` and `cout` to be used in a C++ program. ______________
6. List the preprocessor directive that is used to allow data and output files to be used in the program. ______________
7. Blank spaces or unseen control characters in a data file are referred to as ______________.
8. The << in a `cout` statement is called the ______________ operator.
9. The `#include<`______________`>` is needed for formatted output.
10. The `'\n'` is the escape sequence that ______________.

LESSON 3A

LAB 3.1 Working with the `cin` Statement

Bring in the program `bill.cpp` from the Lab 3 folder. The code is listed below:

```
// This program will read in the qunatity of a particular item and its price.
// It will then print out the total price
// The input will come from the keyboard and the output will go to
// the screen

//   PLACE YOUR NAME HERE

#include <iostream.h>
#include <iomanip.h>
```

continues

```
void main()

{
    int   quantity;           // contains the amount of items purchased
    float itemprice;          // contains the price of each item
    float totalbill;          // contains the total bill.  The price of all items

    cout << setiosflags(ios::fixed) << setiosflags(ios::showpoint) << setprecision(2);
                              //formatted output

    cout << "Please input the number of items bought" << endl;

    // Fill in the input statement that brings in the quantity.

    // Fill in the prompt to ask for the price.

    // Fill in the input statement that brings in the price of each item.

    // Fill in the assignment statement that determines totalbill.

    // Fill in  the output statement that prints a label and total bill to the screen.

}
```

Exercise 1: Complete the program so that a sample run inputting 22 for the number of items bought and 10.98 for the price of each item will produce the results below.

Sample run of the program.

```
Please input the number of items bought
22

Please input the price of each item
10.98

The total bill is $241.56
```

Exercise 2: Once you have the program working, change the instruction:

```
cout << setiosflags(ios::fixed) << setiosflags(ios::showpoint)
     << setprecision(2);
```

to

```
cout << setprecision(2);
```

Rerun the program with the same data given in Exercise 1 above and record your results.

____

What do you think the fixed attribute in the `cout` statement does?

____

Exercise 3: Now change the same instruction discussed in Exercise 2 by making the precision 4. Rerun the program with the same data given in Exercise 1 and record your results.

____

What do you think the `setprecision( )` attribute in the `cout` statement does?

____

The attribute `showpoint` forces all floating point output to show a decimal point even if the values are whole numbers. In some environments this is done automatically.

LAB 3.2 Formatting Output

Look at the following table:

```
PRICE    QUANTITY

 1.95    8
10.89    9
```

Assume that from the left margin, the price takes up fifteen spaces. We could say that the numbers are right justified in a 15-width space. Starting where the price ends the next field (quantity) takes up twelve spaces. We can use the formatted output from Lab 3.1 and the statement `setw(n)` where `n` is some integer to indicate the width to produce such tables.

Bring in the program `tabledata.cpp` from the Lab 3 folder. The code is listed below:

```
// This program will bring in two prices and two quantities of items from the keyboard
// It will print those number in a formatted chart

// PLACE YOUR NAME HERE

#include <iostream.h>
#__________________          // fill in this blank

void main()

{
    float price1, price2;           // The price of 2 items
    int   quantity1, quantity2;     // The quantity of 2 items

    cout << setiosflags(ios::fixed) << setiosflags(ios::showpoint) << setprecision(2);
    cout << "Please input the price and quantity of the first item" << endl;

    // Fill in the input statement that reads in price1 and quantity1 from the keyboard.

    // Fill in the prompt for the second price and quantity.

    // Fill in the input statement that reads in price2 and quantity2 from the keyboard.
```

continues

```
    cout << setw(15) << "PRICE" << setw(12) << "QUANTITY" << endl << endl;

    // Fill in the output statement that prints the first price and quantity.
    // Be sure to use setw() statements

    // Fill in the output statement that prints the second price and quantity.

}
```

Exercise 1: Finish the code above by filling in the blanks and the instructions necessary to execute the following sample run. Note that two or more data items can be input at one time by having at least one blank space between them before hitting the enter key.

```
Please input the price and quantity of the first item
1.95 8

Please input the price and quantity of the second item
10.89 9

          PRICE    QUANTITY

           1.95          8
          10.89          9
```

LAB 3.3 Arithmetic Operations and Math Functions

Bring in the program `righttrig.cpp` from the Lab 3 folder. The code is as follows:

```
// This program will input the value of two sides of a triangle and then determine
// the size of the hypotenuse in order to make it a right triangle

// PLACE YOUR NAME HERE

#include <iostream.h>
#include <math.h>

void main()

{

    float a,b;        // these contain the values of the smaller two sides of the triangle
    float hyp;        // the value of the hypotenuse which is determined by the program

    cout << "Please input the value of the two sides" << endl;
    cin >> a >> b;

    // Fill in the assignment statement that determines the hypotenuse.

    cout << "The sides of the right triangle are " << a << " and " << b << endl;

    cout << "The hypotenuse is " << hyp << endl;

}
```

The formula for finding the hypotenuse is $hyp = \sqrt{a^2 + b^2}$.

How can this be implemented in C++? Hint: You will use two predefined math functions (one of them twice) learned in this lesson. One of them will be "inside" the other.

Exercise 1: Fill in the missing statement so that the following sample run is implemented:

```
Please input the value of the two sides
9 3
The sides of the right triangle are 9 and 3
The hypotenuse is 9.48683
```

Exercise 2: Alter the program so that the sample run now looks like the following:

```
Please input the value of the two sides
9 3
The sides of the right triangle are 9 and 3
The hypotenuse is 9.49
```

Note: This is not a trivial change. You must include another directive as well as use the formatted features discussed in the earlier labs of this lesson. Notice that the change is made only to the value of the hypotenuse and not to the values of 9 and 3.

LESSON 3B

LAB 3.4 Working with Type Casting

Bring in the program `batavg.cpp` from the Lab 3 folder.

Exercise 1: The following is the source code:

```
// This program will determine the batting average of a player
// The number of hits and at bat are set internally in the program

// PLACE YOUR NAME HERE

#include <iostream.h>

const int AT_BAT = 421;
const int HITS = 123;

void main()
{
   int batavg;

   batavg = HITS / AT_BAT;                              // an assignment statement
   cout << "The batting average is " << batavg << endl;   // output the result

}
```

Run this program and record the results. The batting average is ________.

Exercise 2: There is a logical error in this program centering around data types. Does changing the data type of `batavg` from `int` to `float` solve the problem? Make that change and run the program again and record the result.

The batting average is __________.

Exercise 3: Continue to work with this program until you get the correct result. The correct result should be 0.292162.

LAB 3.5 Reading and Writing to a File (Optional)

Bring in `billfile.cpp` from the Lab 3 folder. The code is as follows:

```
// This program will read in the qunatity of a particular item and its price
// It will then print out the total price.
// The input will come from a data file and the output will go to
// an output file.

// PLACE YOUR NAME HERE

#include <fstream.h>
#include <iomanip.h>

void main()
{
    ifstream dataIn;          // defines an input stream for a data file
    ofstream dataOut;         // defines an output stream for an output file
    int quantity;             // contains the amount of items purchased
    float itemprice;          // contains the price of each item
    float  totalbill;         // contains the total bill.  The price of all items

    dataIn.open("transaction.dat");
    dataOut.open("bill.out");

    // Fill in the appropriate code in the blank below
    ________ << setiosflags(ios::fixed) << setiosflags(ios::showpoint) << setprecision(2);
                              //formatted output

    // Fill in the input statement that brings in the quantity and price of the item.

    // Fill in the assignment statement that determines the totalbill.

    // Fill in the output statement that prints the total bill to an output file
    // including a label.

}
```

Exercise 1: Notice that this is an altered version of Lab 3.1. This program gets the information from a file and places the output to a file. You must create the data file. Create a data file called transaction.dat that has the following information:

22

10.98

Exercise 2: Fill in the blank and the statements that will read the data from the file and print the following to `bill.out`:

The total bill is $241.56

LAB 3.6 Student Generated Code Assignments

Option 1: Write a program that will read in 3 grades from the keyboard and will print the average (to 2 decimal places) of those grades to the screen. It should include good prompts and labeled output. Use the examples from the earlier labs to help you. You will want to begin with a design (Lesson Set 1 Pre-lab Reading Assignment gave an introduction for a design similar to this problem). Notice in the sample run that the answer is stored in fixed point notation with two decimal points of precision.

Sample run:

```
Please input the first grade
97

Please input the second grade
98.3

Please input the third grade
95

The average of the three grades is 96.77
```

Option 2: The Woody furniture company sells the following three styles of chairs:

Style	**Price Per Chair**
American Colonial	$ 85.00
Modern	$ 57.50
French Classical	$127.75

Write a program that will input the amount of chairs sold for each style. It will print the total dollar sales of each style as well as the total sales of all chairs in fixed point notation with two decimal places.

Sample run:

```
Please input the number of American Colonial chairs sold
20
Please input the number of Modern chairs sold
15
Please input the number of French Classical chairs sold
5
The total sales of American Colonial chairs $1700.00
The total sales of Modern chairs $862.50
The total sales of French Classical chairs $638.75
The total sales of all chairs $3201.25
```

Option 3: Write a program that will input total sales (sales plus tax) that a business generates for a particular month. The program will also input the state and local sales tax percentage. It will output the total sales plus the state tax and local tax to be paid. The output should be in fixed notation with 2 decimal places.

Sample run:

```
Please input the total sales for the month
1080
Please input the state tax percentage in decimal form (.02 for 2%)
.06
Please input the local tax percentage in decimal form (.02 for 2%)
.02

The total sales for the month is $1080.00
The state tax for the month is $60.00
The local tax for the month is $20.00
```

LESSON SET

4 Conditional Statements

PURPOSE

1. To work with relational operators
2. To work with conditional statements
3. To learn and use nested `if` statements
4. To learn and use logical operators
5. To learn and use the `switch` statement

PROCEDURE

1. Students should read the Pre-lab Reading Assignment before coming to lab.
2. Students should complete the Pre-lab Writing Assignment before coming to lab.

Contents	Pre-requisites	Approximate completion time	Page number	Check when done
Pre-lab Reading Assignment		20 min.	36	
Pre-lab Writing Assignment	Pre-lab reading	10 min.	41	
LESSON 4A				
Lab 4.1 Working with Relational Operators and the `if` Statement	Basic understanding of relational operators and the simple `if` statement	15 min.	41	
Lab 4.2 `if/else` and Nested `if` Statements	Basic understanding of nested `if` statements	20 min.	42	
Lab 4.3 Logical Operators	Basic understanding of logical operators	15 min.	43	
LESSON 4B				
Lab 4.4 The `switch` Statement	Understanding of the `switch` statement	25 min.	45	
Lab 4.5 Student Generated Code Assignments	Basic understanding of conditional statements	30 min.	46	

PRE-LAB READING ASSIGNMENT

Relational Operators

You have already seen that the statement `total = 5` is an assignment statement; that is, the integer 5 is placed in the variable called `total`. Nothing relevant to our everyday understanding of equality is present here. So how do we deal with equality in a program? How about greater than or less than? C++ allows the programmer to compare numeric values using **relational operators**. They are the following:

>	Greater than
<	Less than
> =	Greater than or equal to
< =	Less than or equal to
= =	Equal to
! =	Not equal to

An expression of the form `num1 > num2` is called a **relational expression**. Note that it does *not* assert that `num1` is greater than `num2`. It actually tests to see if this is true. So relational expressions are boolean. Their value must be either *true* or *false*. The statement `cost!=9` is false if cost has value 9 and true otherwise. Consider the following code:

```
int years;
years = 6;
years == 5;
years = years - 1;
years == 5;
```

In this sequence the first occurrence of `years == 5` is a false statement whereas the second occurrence is true. Can you see why?

The `if` Statement

Sometimes we may only want a portion of code executed under certain conditions. To do so, we use **conditional statements**. For example, if you are writing a payroll program to compute wages, then the program should only compute overtime pay *if* the employee worked more than 40 hours in a given week. Otherwise, when the program is executed the overtime portion of the code should be bypassed. An **`if` statement** is one kind of conditional statement.

Consider the following program:

Sample Program 4.1:

```
// This program prints "You Pass" if a student's average is above 60 or prints
// "You Fail" otherwise

#include <iostream.h>

void main()
{
     float average;
```

```
    cout << "Input your average" << endl;
    cin >> average;

    if (average >= 60)   // note the use of a relational operator
        cout << "You Pass" << endl;

    if (average < 60)
        cout << "You Fail" << endl;

}
```

Note that it is not possible for this program to print out both "You Pass" and "You Fail". Only one of the `if` statements will be executed. Later we will see a way to write this program without using 2 `if` statements.

If you want to conditionally execute several statements using `if`, the following syntax is required:

```
if (expression)
{
    statement_1;
    statement_2;
        :
    statement_n;
}
```

Note the curly braces surrounding the set of statements to be conditionally executed.

The `if/else` Statement

In Sample Program 4.1 we used two `if` statements. A more elegant approach would be to use the **`if/else` statement** as follows:

```
if (average >= 60)
    cout << "You Pass" << endl;

else
    cout << "You Fail" << endl;
```

In every `if/else` statement the program can take only one of two possible paths. Multiple statements can be handled using curly braces in the same way as the `if` statement.

The `if/else if` Statement

The `if/else` statement works well if there are only two possible paths to follow. However, what if there are more than two possibilities. For example, suppose we need to decide what kind of vacation to take based on a yearly work bonus:

if the bonus is less than $1,000, we set up a tent and eat spam in the back yard

if the bonus is less than $10,000 and greater than or equal to $1,000, we go to Disney World

if the bonus is $10,000, we go to Hawaii

We could code this using the **if/else if** statement as follows:

```
float bonus;

cout << "Please input the amount of your yearly bonus" << endl;
cin >> bonus;

if (bonus < 1000)
   cout << "Another vacation eating spam on the lawn" << endl;

else if (bonus < 10000)
   cout << "Off to Disney World!" << endl;

else if (bonus == 10000)
   cout << "Lets go to Hawaii!" << endl;
```

Can you explain why the first `else if` conditional statement does not require a greater than or equal to 1000 condition?

In general we can use as many `else if` expressions as needed to solve a given problem.

The Trailing `else`

What happens in the code above if the bonus entered is greater than $10,000? Actually, nothing will happen since none of the conditional expressions are true in this case. Sometimes it is advantageous to add a final or **trailing `else`** at the end of a chain of `if/else if` statements to handle "all other cases." For example, we could modify the code to read:

```
if (bonus < 1000)
    cout << "Another vacation on the lawn" << endl;
else if (bonus < 10000)
    cout << "Off to Disney World!" << endl;
else if (bonus == 10000)
    cout << "Lets go to Hawaii!" << endl;
else
{
    cout << bonus << " is not a valid bonus" << endl;
    cout << "Please run the program again with valid data" << endl;
}   // note the necessary use of the curly brackets here
```

Of course, few would complain about a bonus greater than $10,000 and the Hawaii trip could still be done on this budget. However, if the maximum possible bonus is $10,000, then the trailing `else` will let the user know that an illegal value has been entered.

Nested `if` Statements

Often programmers use an `if` statement within another `if` statement. For example, suppose a software engineering company wants to screen applicants first for experienced programmers and second for C++ programmers specifically. One possible program is the following:

Sample Program 4.2:

```
#include <iostream.h>

void main()
{
     char Programmer, CplusPlus;

     cout << "Before we consider your application, answer the following"
          << endl;
     cout << " yes ( enter Y ) or no ( enter N )" << endl;
     cout << "Are you a computer programmer?" << endl;

     cin >> Programmer;

     if (Programmer == 'Y')
     {
         cout << "Do you program in C++?" << endl;
         cin >> CplusPlus;

         if (CPlusPlus == 'Y')
            cout << " You look like a promising candidate for employment"
                 << endl;
         else if (CPlusPlus == 'N')
            cout << " You need to learn C++ before further consideration"
                 << endl;
         else
            cout << " You must enter Y or N" << endl;
     }

    else if (Programmer == 'N')
         cout << " You are not currently qualified for employment" << endl;

    else
         cout <<  " You must enter Y or N" << endl;

}
```

Note how C++ programmers are identified using a nested `if` statement. Also note how the trailing `else` is used to detect invalid input.

Logical Operators

By using relational operators C++ programmers can create relational expressions. Programmers can also reverse the "truth value" of a single expression using **logical operators**. For example, instead of a statement such as "if it is sunny, then we will go outside," one may use a statement such as "if it is sunny and it is warm, then we will go outside." Note that this statement has two smaller statements "it is sunny" and "it is warm" joined by the **AND** logical operator. To satisfy the `if` conditional, both the sunny and warm requirements must be met.

The **NOT** operator negates a single statement. For example, "it is sunny" can be negated by "it is not sunny."

The **OR** operator is similar to the AND in that it connects two statements. However, there is an ambiguity about the meaning of the word *or* in English. In the statement "tonight at 8:00 I will go to the concert in the park or I will go to the stadium to see the ball game," the word **or** is *exclusive*. That is, I can go to the concert or to the game but not both. However, in the statement "I need to draw an ace or a king to have a good poker hand," the word **or** is *inclusive*. In other words, I can draw a king, an ace, or even both, and I will have a good hand. So we have a choice to make. Let A and B be two statements. A OR B could mean A or B but not both. It could also mean A or B or both. In computer science we use the second meaning of the word *or*. For example, in the statement "if it is sunny or it is warm, then I will go outside," there are three scenarios where I will go outside: if it is sunny but not warm, if it is warm but not sunny, or if it is sunny and warm.

The syntax used by C++ for logical operators is the following:

AND	&&
OR	||
NOT	!

Consider the following:

```
if (dollars <= 0 || !(accountActive) )
    cout << " You may not withdraw money from the bank";
```

It is good programming practice to enclose the operand after the (!) operator in parentheses. Unexpected things can happen in complicated expressions if you do not. When will this code execute the `cout` statement? What type of variable do you think `accountActive` is?

The `switch` Statement

We have already seen how `if` statements can affect the branching of a program during execution. Another way to do this is using the **`switch` statement**. It is also a conditional statement. The `switch` statement uses the value of an integer expression to determine which group of statements to branch through. The sample program below illustrates the syntax.

Sample Program 4.3:

```
#include <iostream.h>

void main()
{
     char grade;

     cout << "What grade did you earn in Programming I?" << endl;
     cin >> grade;

     switch( grade )     // This is where the switch statement begins
      {
          case 'A':cout << "an A - excellent work!" << endl;
                break;
```

```
            case 'B':cout << "you got a B - good job" << endl;
                  break;
            case 'C':cout << "earning a C is satisfactory" << endl;
                  break;
            case 'D':cout << "while D is passing, there is a problem" << endl;
                  break;
            case 'F':cout << "you failed - better luck next time" << endl;
                  break;
            default:cout << "You did not enter an A, B, C, D, or F" << endl;
      }
}
```

Note the use of the curly braces that enclose the cases. Also, consider the variable `grade`. It is declared as a character data type and the case statements have character arguments such as `'B'`. This seems to contradict what we said above, namely that the `switch` statement uses the value of integer expressions to determine branching. However, this apparent contradiction is resolved by the compiler automatically converting character data into the integer data type. Finally, notice the role of the `default` statement. The `default` branch is followed if none of the case expressions match the given `switch` expression.

PRE-LAB WRITING ASSIGNMENT

Fill-in-the-Blank Questions

1. The two possible values for a relational expression are _______________ or _______________.
2. C++ uses the _______________ symbol to represent the AND operator.
3. The `switch` statement and `if` statements are examples of _______________ statements.
4. In C++ is the meaning of the OR logical operator inclusive or exclusive? _______________
5. C++ uses the _______________ symbol to represent the OR operator.
6. It is good programming practice to do what to the operand after the NOT operator? _______________
7. The `switch` statement uses the value of an _______________ expression to determine which group of statements to branch through.
8. In a `switch` statement the _______________ branch is followed if none of the case expressions match the given `switch` expression.
9. C++ allows the programmer to compare numeric values using _______________.
10. The C++ symbol for equality is _______________.

LESSON 4A

LAB 4.1 Working with Relational Operators and the `if` Statement

Exercise 1: Bring in the file `initialize.cpp` from the Lab 4 folder. Run the program several times using a different input each time. The code is on the following page.

```
// This program tests whether or not an initialized value is equal to a value
// input by the user

// PLACE YOUR NAME HERE

#include <iostream.h>

void main( )
{
   int num1, num2 = 5;              // note that num2 has been initialized to 5

   cout << "Please enter an integer" << endl;
   cin >> num1;


   if (num1 = num2)
   cout << "Hey, thats  a coincidence!" << endl;

   if (num1 != num2 )
   cout << "The values are not the same" << endl;

}
```

Exercise 2: Does this program do what you expect? If so, explain what it is doing. If not, explain how the program can be fixed.

Exercise 3: Adjust the program so that the user inputs both values and tests to see if they are equal. Also output the values of `num1` and `num2` by adding code inside each `if` statement. What happens if you forget the curly braces in the `if` statements?

LAB 4.2 `if/else` and Nested `if` Statements

Exercise 1: Bring in the file `condition.cpp` from the Lab 4 folder. Adjust it so that it uses an `if/else` statement rather than two `if` statements.

```
// This program tests whether or not an initialized value is equal to a value
// input by the user

// PLACE YOUR NAME HERE

#include <iostream.h>

void main( )
{
   int num1, num2 = 5;              // note that num2 has been initialized to 5

   cout << "Please enter an integer" << endl;
   cin >> num1;

   if (num1 == num2)
   {
```

```
        cout << "Hey, thats a coincidence!" << endl;
        cout << "num1= " << num1 << endl;
        cout << "num2= " << num2 << endl;
    }

    if (num1 != num2 )
    {
        cout << "The values are not the same" << endl;
        cout << "num1= " << num1 << endl;
        cout << "num2= " << num2 << endl;
    }

}
```

Exercise 2: Repeat Exercise 3 from Lab 4.1 with your `if/else` version from this lab.

Exercise 3: Retrieve `grade.cpp` from the Lab 4 folder and alter it using the `if/else if` statement so that it has the following categories: Invalid data (data above 100), "A" category (90–100), "B" category (80–89), "You Pass" category (60–79), "You Fail" category (0–59). What will happen to your program if you enter a negative value such as -12?

The code is as follows:

```
//  This program prints "You Pass" if a student's average is above 60
//  or prints "You Fail" otherwise

// PLACE YOUR NAME HERE

#include <iostream.h>

void main()
{

    float average;     //holds the grade average

    cout << "input your average" << endl;
    cin >> average;

    if (average >= 60)
        cout << "You Pass" << endl;

    if (average < 60)
        cout << "You Fail" << endl;

}
```

Exercise 4: Alter the program in Exercise 3 so that it will have an invalid number category that will include any negative number as well as numbers over 100.

LAB 4.3 Logical Operators

Retrieve `LogicalOp.cpp` from the Lab 4 folder. The code is as follows:

```
// This program illustrates the use of logical operators

// PLACE YOUR NAME HERE

#include <iostream.h>

void main()
{
   char Year;
   float GPA;

   cout << "What year student are you ?" << endl;
   cout << "Enter 1 (freshman), 2 (sophomore), 3 (junior), or 4 (senior)"
        << endl << endl;
   cin >> Year;

   cout << "Now enter your GPA" << endl;
   cin >> GPA;

   if (GPA >= 2.0 && Year == '4')
      cout << "It is time to graduate soon" << endl;

   else if (Year != '4'|| GPA <2.0)
      cout << "You need more schooling" << endl;

}
```

Exercise 1: How could you rewrite `GPA >= 2.0` in the first `if` statement using the NOT operator?

Exercise 2: Could you replace `Year !='4'` in the `else if` statement with `Year < 4` or `Year <= 3`? Why or why not?

Exercise 3: If you replace

```
if ( GPA >= 2.0 && Year == '4')
```

with

```
if ( GPA >= 2.0 || Year == '4')
```

and replace

```
else if ( Year != '4'|| GPA < 2.0)
```

with

```
else if ( Year != '4'&& GPA < 2.0)
```

which students will graduate and which will not graduate according to this new program?

Does this handle all cases (i.e., all combinations of `Year` and `GPA`)?

Exercise 4: Could you replace `else if ( Year != '4'|| GPA < 2.0)` with the single word `else`?

LESSON 4B

LAB 4.4 The `switch` Statement

Exercise 1: Bring in the file `switch.cpp` from the Lab 4 folder. This is Sample Program 4.3 from the Pre-lab Reading Assignment. The code is shown below. Remove the break statements from each of the cases. What is the effect on the execution of the program?

```
// PLACE YOUR NAME HERE

#include <iostream.h>

void main()
{
    char grade;

    cout << "What grade did you earn in Programming I ?" << endl;
    cin >> grade;

    switch( grade )              // This is where the switch statement begins
    {
        case 'A':cout << "an A - excellent work !" << endl;
             break;
        case 'B':cout << "you got a B - good job" << endl;
             break;
        case 'C':cout << "earning a C is satisfactory" << endl;
             break;
        case 'D':cout << "while  D is passing, there is a problem" << endl;
             break;
        case 'F':cout << "you failed - better luck next time" << endl;
             break;
        default:cout << "You did not enter an A, B, C, D, or F" << endl;
    }

}
```

Exercise 2: Add an additional case statement that allows for a Passing option for a grade of D or better. Use the sample run given below to model your output.

Sample Run:

```
What grade did you earn in Programming I ?
A
YOU PASSED!
an A - excellent work!
```

Exercise 3: Rewrite the program `switch.cpp` using `if` and `else if` statements rather than a `switch` statement. Did you use a trailing `else` in your new version? If so, what did it correspond to in the original program with the `switch` statement?

LAB 4.5 Student Generated Code Assignments

Option 1: Write a program that prompts the user for their quarterly water bill for the last four *quarters*. The program should find and output their average *monthly* water bill. If the average bill exceeds $75, the output should include a message indicating that too much water is being used. If the average bill is at least $25 but no more than $75, the output should indicate that a typical amount of water is being used. Finally, if the average bill is less than $25, the output should contain a message praising the user for conserving water. Use the sample run below as a model for your output.

Sample Run 1:

```
Please input your water bill for quarter 1:
300

Please input your water bill for quarter 2:
200

Please input your water bill for quarter 3:
225

Please input your water bill for quarter 4:
275

Your average monthly bill is $83.33. You are using excessive amounts of water.
```

Sample Run 2:

```
Please input your water bill for quarter 1:
100

Please input your water bill for quarter 2:
150

Please input your water bill for quarter 3:
75

Please input your water bill for quarter 4:
125

Your average monthly bill is $37.50. You are using a typical amount of water.
```

Option 2: The local t-shirt shop sells shirts that retail for $12. Quantity discounts are given as follow:

Number of Shirts	Discount
5–10	10%
11–20	15%
21–30	20%
31 or more	25%

Write a program that prompts the user for the number of shirts required and then computes the total price. Make sure the program accepts only nonnegative input.

Use the following sample runs to guide you:

Sample Run 1:

```
How many shirts would you like ?
4

The cost per shirt is $12 and the total cost is $48.
```

Sample Run 2:

```
How many shirts would you like ?
0

The cost per shirt is $12 and the total cost is $0.
```

Sample Run 3:

```
How many shirts would you like ?
8

The cost per shirt is $10.80 and the total cost is $86.40.
```

Sample Run 4:

```
How many shirts would you like ?
-2

Invalid Input: Please enter a nonnegative integer.
```

Option 3: The University of Guiness charges $3000 per semester for in-state tuition and $4500 per semester for out-of-state tuition. In addition, room and board is $2500 per semester for in-state students and $3500 per semester for out-of-state students. Write a program that prompts the user for their residential status (i.e., in-state or out-of-state) and whether they require room and board (Y or N). The program should then compute and output their bill for that semester.

Use the sample output below:

Sample Run 1:

```
Please input "I" if you are in-state or "O" if you are out-of-state:
I

Please input "Y" if you require room and board and "N" if you do not:
N

Your total bill for this semester is $3000.
```

Sample Run 2:

```
Please input "I" if you are in-state or "O" if you are out-of-state:
O

Please input "Y" if you require room and board and "N" if you do not:
Y

Your total bill for this semester is $8000.
```

LESSON SET

5 Looping Statements

PURPOSE

1. To introduce counter and event controlled loops
2. To work with the `while` loop
3. To introduce the `do-while` loop
4. To work with the `for` loop
5. To work with nested loops

PROCEDURE

1. **Students should read the Pre-lab Reading Assignment before coming to lab.**
2. **Students should complete the Pre-lab Writing Assignment before coming to lab.**
3. **In the lab, students should complete labs assigned to them by the instructor.**

Contents	**Pre-requisites**	**Approximate completion time**	**Page number**	**Check when done**
Pre-lab Reading Assignment		20 min.	50	
Pre-lab Writing Assignment	Pre-lab reading	10 min.	56	
LESSON 5A				
Lab 5.1 Working with the `while` Loop	Basic understanding of the `while` loop	25 min.	56	
Lab 5.2 Working with the `do-while` Loop	Basic understanding of `do-while` loop	25 min.	58	
LESSON 5B				
Lab 5.3 Working with the `for` Loop	Understanding of `for` loops	15 min.	59	
Lab 5.4 Nested Loops	Understanding of nested `for` loops	15 min	60	
Lab 5.5 Student Generated Code Assignments	Basic understanding of loop control structures	30 min.	62	

PRE-LAB READING ASSIGNMENT

Increment and Decrement Operator

To execute many algorithms we need to be able to add or subtract 1 from a given integer quantity. For example:

```
count = count + 1;      // what would happen if we used ==
                        // instead of = ?
count += 1;
```

Both of these statements **increment** the value of `count` by 1. If we replace "+" with "-" in the above code, then both statements **decrement** the value of `count` by 1. C++ also provides an **increment operator** ++ and a **decrement operator** -- to perform these tasks. There are two modes that can be used:

```
count++;     // increment operator in the postfix mode
count--;     // decrement operator in the postfix mode

++count;     // increment operator in the prefix mode
--count;     // decrement operator in the prefix mode
```

The two increment statements both execute exactly the same. So do the decrement operators. What is the purpose of having postfix and prefix modes? To answer this, consider the following code:

```
int age = 49;
if (age++ > 49)
    cout << "Congratulations - You have made it to the half-century"
         << " mark !" << endl;
```

In this code, the `cout` statement will not execute. The reason is that in the postfix mode the comparison between `age` and 49 is made *first*. Then the value of `age` is incremented by one. Since 49 is not greater than 49, the `if` conditional is false. Things are much different if we replace the postfix operator with the prefix operator:

```
int age = 49;
if (++age > 49)
    cout << " Congratulations - You have made it to the half-century"
         << " mark !" << endl;
```

In this code `age` is incremented first. So its value is 50 when the comparison is made. The conditional statement is true and the `cout` statement is executed.

The `while` Loop

Often in programming one needs a statement or block of statements to repeat during execution. This can be accomplished using a **loop**. A loop is a control structure that causes repetition of code within a program. C++ has three types of loops. The first we will consider is the **`while` loop**. The syntax is the following:

```
while (expression)
{
      statement_1;
      statement_2;
          :
      statement_n;
}
```

If there is only one statement, then the curly braces can be omitted. When a `while` loop is encountered during execution, the expression is tested to see if it is true or false. The block of statements are repeated as long as the expression is true. Consider the following:

Sample Program 5.1:

```
#include <iostream.h>

void main()
{
      int num = 5;
      int numfac = 1;

      while (num > 0)
              numfac = numfac*num-- ;       // note the use of the decrement
                                            // operator in the postfix mode
      cout << " 5! =" << numfac << endl;

}
```

This program computes 5!=5*4*3*2*1 and then prints the result to the screen. Note how the `while` loop controls the execution. Since `num = 5` when the `while` loop is first encountered, the block of statements in the body of the loop is executed at least once. In fact, the block is executed 5 times because of the decrement operator which forces the value of `num` to decrease by one every time the block is executed. When `num=0` the expression `num > 0` is false and so the loop is exited. Then the `cout` statement is executed.

What do you think will happen if we replace `numfac*num--` with just `numfac*num` in the above code? Without the decrement operator, `num` always has a value of 5. This means that the expression `num > 0` is always true! If we try to execute the modified program, the result is an **infinite loop**, i.e., a block of code that will repeat forever. One must be very cautious when using loops to ensure that the loop will terminate. Here is another example where the user may have troubles with termination.

Sample Program 5.2:

```
#include <iostream.h>

void main()
{
      char letter = 'a';

      while (letter != 'x')
      {
              cout << "Please enter a letter" << endl;
              cin >> letter;
              cout << "The letter you entered is " << letter << endl;
      }

}
```

Note that this program requires input from the user during execution. Infinite loops can be avoided, but it would help if the user knew that the `'x'` character terminates the execution. Without this knowledge the user could continually enter characters other than `'x'` and never realize how to terminate the program. In the lab assignments you will be asked to modify this program to make it more user friendly.

Counters

Often a programmer needs to keep track of the number of times a particular loop is repeated. This can be done using a **counter**, i.e., a variable that is incremented or decremented each time a loop repeats. Sample Program 5.3 computes the test average for scores entered by the user. A counter is used to keep track of the number of test scores that the user enters. Execution is controlled by a `while` loop that terminates when the user enters the **sentinel value** –1. A sentinel is used to indicate the end of a list of values. We chose –1 for this value since it is an invalid test score. It would not make sense to use a sentinel value between 0 and 100 since this is the range of valid test scores.

Sample Program 5.3:

```
#include <iostream.h>

void main()
{
     int score = 0;
     float total = 0.0;
     float average = 0.0;
     int counter = 0;

     while (score != - 1)
     {
            total = total + score;
            cout << " Enter your score on test " << ++counter << endl
                 << " To exit, enter -1 " << endl;
            cin >> score;

     }

     average = total / (counter-1);

     cout << " Your average based on " << (counter-1) << " test scores is "
          << average << endl;

}
```

Why was `counter-1` used to compute the average rather than `counter`?

The `do-while` Loop

The `while` loop is a **top test** loop. Since we test the expression before entering the loop, if the test expression in the `while` loop is initially false, then no iterations of the loop will be executed. If the programmer wants the loop to be executed at least once, then a **bottom test loop** should be used. C++ provides the **`do-while` loop** for this purpose. A `do-while` loop is similar to a `while` loop except that the statements inside the loop body are executed *before* the expression is tested. The format for a single statement in the loop body is the following:

```
do
   statement;
while (expression);
```

Note that the statement must be executed once even if the expression is false. To see the difference between these two loops consider the code

```
int num1 = 5;
int num2 = 7;

while (num2 < num1)
{
        num1 = num1 + 1;
        num2 = num2 - 1;
}
```

Here the statements `num1 = num1 + 1` and `num2 = num2 - 1` are never executed since the test expression `num2 < num1` is initially false. However, we get a different result using a `do-while` loop:

```
int num1 = 5;
int num2 = 7;

do
{
        num1 = num1 + 1;
        num2 = num2 - 1;
} while (num2 < num1);
```

In this code the statements `num1 = num1 + 1` and `num2 = num2 - 1` are executed exactly once. At this point `num1 = 6` and `num2 = 6` so the expression `num2 < num1` is false. Consequently, the program exits the loop and moves to the next section of code. Also note that since we need a block of statements in the loop body, curly braces must be placed around the statements. In Lab 5.2 you will see how `do-while` loops can be useful for programs that involve a repeating menu.

Both `while` and `do-while` loops are examples of **event controlled loops**. In the next section we will see a **counter controlled loop**.

The `for` Loop

The **`for` loop** is often used for applications that require a counter. For example, suppose we want to find the average (mean) of the first n positive integers. By definition, this means that we need to add $1 + 2 + 3 + \ldots + n$ and then divide by n. Note this should just give us the value in the "middle" of the list $1, 2, \ldots, n$. Since we know exactly how many times we are performing a sum, the `for` loop is the natural choice.

The syntax for the `for` loop is the following:

```
for (initialization; test; update)
{
      statement_1;
      statement_2;
          :
      statement_n;
}
```

Now we are ready to add the first *n* positive integers and find their mean value.

Sample Program 5.4:

```
#include <iostream.h>

void main()
{
      int value;
      int total = 0;
      int number;
      float mean;

      cout << "Please enter a positive integer" << endl;
      cin >> value;

      if (value > 0)
      {
            for (number = 1; number <= value; number++)
            {
                  total = total + number;
            }                                // curly braces are optional since
                                             // there is only one statement

            mean = float(total) / value;     // note the use of the typecast
                                             // operator

            cout << "The mean average of the first " << value
                  << " positive integers is " << mean << endl;
      }
      else
            cout << "Invalid input - integer must be positive" << endl;

}
```

Note that the counter in the `for` loop of Sample Program 5.4 is `number`. It increments from 1 to `value` during execution. There are several other features of this code that also need to be addressed. First of all, why is the typecast operator needed to compute the mean? What do you think will happen if it is removed? Finally, what would happen if we entered a float such as 2.99 instead of an integer? Lab 5.3 will demonstrate what happens in these cases.

Nested Loops

Often programmers need to use a loop within a loop, or **nested loops**. Sample Program 5.5 below provides a simple example of a nested loop. This program finds the average number of hours per day spent programming by each student over a three-day weekend. The outer loop controls the number of students and the inner loop allows the user to enter the number of hours worked each of the three days for a given student. Note that the inner loop is executed three times for each iteration of the outer loop.

Sample Program 5.5:

```
// This program finds the average time spent programming by a student each
// day over a three day period.

#include <iostream.h>

void main()
{
     int numStudents;
     float numHours, total, average;
     int count1 = 0, count2 = 0;      // these are the counters for the loops

     cout << "This program will find the average number of hours a day"
          << " that each given student spent programming over a long weekend"
          << endl << endl;

     cout << "How many students are there ?" << endl << endl;
     cin >> numStudents;

     for (count1 = 1; count1 <= numStudents; count1++)
     {
          total = 0;
          for (count2 = 1; count2 <= 3; count2++)
          {
               cout << "Please enter the number of hours worked by student "
                    << count1 << " on day " << count2 << "." << endl;
               cin >> numHours;

               total = total + numHours;
          }
          average = total / 3;

          cout << endl;
          cout << "The average number of hours per day spent programming by"
               << " student " << count1 <<" is " << average
               << endl << endl << endl;
     }
}
```

In Lab 5.4 you will be asked to modify this program to make it more flexible.

PRE-LAB WRITING ASSIGNMENT

Fill-in-the-Blank Questions

1. A block of code that repeats forever is called ______________.
2. To keep track of the number of times a particular loop is repeated, one can use a ______________.
3. An event controlled loop that is always executed at least once is the ______________.
4. An event controlled loop that is not guaranteed to execute at least once is the ______________.
5. In the conditional `if(++number < 9)`, the comparison `number < 9` is made ______________ and `number` is incremented ______________. (Choose first or second for each blank.)
6. In the conditional `if(number++ < 9)`, the comparison `number < 9` is made ______________ and `number` is incremented ______________. (Choose first or second for each blank.)
7. A loop within a loop is called a ______________.
8. To write out the first 12 positive integers and their cubes, one should use a ______________.
9. A ______________ value is used to indicate the end of a list of values. It can be used to control a `while` loop.
10. In a nested loop the ______________ loop goes through all of its iterations for each iteration of the ______________ loop. (Choose inner or outer for each blank.)

LESSON 5A

LAB 5.1 Working with the `while` Loop

Bring in program `while.cpp` from the Lab 5 folder. (This is Sample Program 5.2 from the Pre-lab Reading Assignment). The code is shown below:

```
// PLACE YOUR NAME HERE

#include <iostream.h>

void main()
{
   char letter = 'a';

   while (letter != 'x')
   {
         cout << "Please enter a letter" << endl;
         cin >> letter;
         cout << "The letter you entered is " << letter << endl;
   }

}
```

Exercise 1: This program is not user friendly. Run it a few times and explain why.

Exercise 2: Add to the code so that the program is more user friendly.

Exercise 3: How would this code affect the execution of the program if the `while` loop is replaced by a `do-while` loop? Try it and see.

Bring in program `sentinel.cpp` from the Lab 5 Folder. The code is shown below:

```
// This program illustrates the use of a sentinel in a while loop.
// The user is asked for monthly rainfall totals until a sentinel
// value of -1 is entered. Then the total rainfall is displayed

// PLACE YOUR NAME HERE

#include <iostream.h>

void main()
{
   // Fill in the code to declare and initialize to 1 the variable count
   float total = 0, rain;

   cout << "Enter the total rainfall for month " << count << endl;
   cout << "Enter -1 when you are finished" << endl;
   // Fill in the code to read in the value for rain

   // Fill in the code to start a while loop that iterates while rain does not equal -1
   {
      // Fill in the code to update total by adding it to rain
      // Fill in the code to increment count by one

      cout << "Enter the total rainfall in inches for month "
           << count << endl;
      cout << "Enter -1 when you are finished" << endl;
      // Fill in the code to read in the value for rain

   }

   if (count == 1)
      cout << "No data has been entered" << endl;

   else
      cout << "The total rainfall for the " << count-1 << " months is "
           << total << " inches." << endl;

}
```

Exercise 4: Complete the program above by filling in the code described in the statements in bold so that it will perform the indicated task.

Exercise 5: Run the program several times with various input. Record your results. Are they correct? What happens if you enter –1 first? What happens if you enter only values of 0 for one or more months? Is there any numerical data that you should not enter?

Exercise 6: What is the purpose of the following code in the program above?

```
if (count == 1)
    cout << "No data has been entered" << endl;
```

LAB 5.2 Working with the `do-while` Loop

Bring in the program `dowhile.cpp` from the Lab 5 folder. The code is shown below:

```
// This program displays a hot beverage menu and prompts the user to
// make a selection. A switch statement determines which item the user
// has chosen. A do-while loop repeats until the user selects item E
// from the menu.

// PLACE YOUR NAME HERE

#include <iostream.h>
#include <iomanip.h>

void main()
{
    // Fill in the code to declare an integer variable number, a floating
    // point variable cost, and a character variable beverage

    cout << setiosflags(ios::fixed) << setiosflags(ios::showpoint) << setprecision(2);

    do
    {
        cout << endl << endl;
        cout << "Hot Beverage Menu" << endl << endl;
        cout << "A: Coffee" << endl;
        cout << "B: Tea" << endl;
        cout << "C: Hot Chocolate" << endl;
        cout << "D: Cappuccino" << endl <<endl << endl;

        cout << "Enter the beverage A,B,C, or D you desire" << endl;
        cout << "Enter E to exit the program" << endl << endl;
        // Fill in the code to read in beverage

        if(beverage != 'E')
        {
            cout << "How many cups would you like ?" << endl;
            //  Fill in the code to read in number
        }
```

```
        // Fill in the code to develop a switch statement that is controlled by beverage
        {
        case 'A': cost = number * 1.0;
               cout << "The total cost is $ " << cost << endl;
               break;
        // Fill in the code to give the case for tea with cost $0.75 per cup
        // Fill in the code to give the case for hot chocolate with cost $1.25 per cup
        // Fill in the code to give the case for cappuccino with cost $2.50 per cup

        case 'E': cout << " Please come again" << endl;
               break;
        default:cout << // Fill in the code to write a message indicating invalid
                        // selections
               cout << " Try again please" << endl;
        }
    } // Fill in the code to finish the do-while statement with the condition that
      // beverage does not equal E.

}
```

Exercise 1: Complete the program above by filling in the statements in bold so that it performs the indicated task.

Exercise 2: Test your program with different input. Try all the possible relevant cases. Record your results. What happens if you do not enter A, B, C, D, or E? Is there any way the program could be improved when invalid input is entered?

LESSON 5B

LAB 5.3 Working with the `for` Loop

Bring in program `for.cpp` from the Lab 5 folder (this is Sample Program 5.4 from the Pre-lab Reading Assignment). This program has the user input a number n and then finds the mean of the first n positive integers. The code is shown below:

```
// PLACE YOUR NAME HERE

#include <iostream.h>


void main()
{
    int value;
    int total = 0;
    int number;
    float mean;

    cout << "Please enter a positive integer" << endl;
    cin >> value;

    if (value > 0)
```

continues

```
   {
      for (number = 1; number <= value; number++)
      {
         total = total + number;
      }           // curly braces are optional since there is only one statement

      mean = float(total) / value;               // note the use of the typecast
                                                 // operator here
      cout << "The mean average of the first " << value
           << " positive integers is " << mean << endl;
   }
   else
      cout << "Invalid input - integer must be positive" << endl;

}
```

Exercise 1: Why is the typecast operator needed to compute the mean in the statement `mean = float(total)/value;`? What do you think will happen if it is removed? Modify the code and try it. Record what happens. Make sure that you try both even and odd cases. Now put `float(total)` back in the program.

Exercise 2: What happens if you enter a float such as 2.99 instead of an integer for `value`? Try it and record the results.

Exercise 3: Modify the code so that it computes the mean of the consecutive positive integers $n, n+1, n+2, \ldots, m$, where the user chooses n and m. For example, if the user picks 3 and 9, then the program should find the mean of 3, 4, 5, 6, 7, 8, and 9, which is 6.

LAB 5.4 Nested Loops

Bring in program `nested.cpp` from the Lab 5 folder (this is Sample Program 5.5 from the Pre-lab Reading Assignment). The code is shown below:

```
// This program finds the average time spent programming by a student each day
// over a three day period.

// PLACE YOUR NAME HERE

#include <iostream.h>

void main()
{
   int numStudents;
   float numHours, total, average;
   int count1 = 0, count2 = 0;          // these are the counters for the loops

   cout << "This program will find the average number of hours a day"
        << " that a student spent programming over a long weekend" << endl << endl;
   cout << "How many students are there ?" << endl << endl;
   cin >> numStudents;
```

```
    for(count1 = 1; count1 <= numStudents; count1++)
    {
        total = 0;
        for(count2 = 1; count2 <= 3; count2++)
        {
            cout << "Please enter the number of hours worked by student " << count1
                 << " on day " << count2 << "." << endl;
            cin >> numHours;

            total = total + numHours;
        }

        average = total / 3;

        cout << endl;
        cout << "The average number of hours per day spent programming by "
             << "student " << count1 << " is " << average
             << endl << endl << endl;
    }
}
```

Exercise 1: Note that the inner loop of this program is always executed exactly three times—once for each day of the long weekend. Modify the code so that the inner loop iterates *n* times, where *n* is a positive integer input by the user. In other words, let the user decide how many days to consider just as they choose how many students to consider.

Sample Run:

```
This program will find the average number of hours a day that a student spent
programming over a long weekend

How many students are there?
2
Enter the number of days in the long weekend
2

Please enter the number of hours worked by student 1 on day 1
4

Please enter the number of hours worked by student 1 on day 2
6

The average number of hours per day spent programming by student 1 is 5


Please enter the number of hours worked by student 2 on day 1
9

Please enter the number of hours worked by student 2 on day 2
13

The average number of hours per day spent programming by student 2 is 11
```

Exercise 2: Modify the program from Exercise 1 so that it also finds the average number of hours per day that a given student studies biology as well as programming. For each given student include two prompts, one for each subject. Have the program print out which subject the student, on average, spent the most time on.

LAB 5.5 Student Generated Code Assignments

Option 1: Write a program that performs a survey tally on beverages. The program should prompt for the next person until a sentinel value of –1 is entered to terminate the program. Each person participating in the survey should choose their favorite beverage from the following list:

1. Coffee 2. Tea 3. Coke 4. Orange Juice

Sample Run:

```
Please input the favorite beverage of person #1: Choose 1, 2, 3, or 4 from the
above menu or -1 to exit the program
4

Please input the favorite beverage of person #2: Choose 1, 2, 3, or 4 from the
above menu or -1 to exit the program
1

Please input the favorite beverage of person #3: Choose 1, 2, 3, or 4 from the
above menu or -1 to exit the program
3

Please input the favorite beverage of person #4: Choose 1, 2, 3, or 4 from the
above menu or -1 to exit the program
1

Please input the favorite beverage of person #5: Choose 1, 2, 3, or 4 from the
above menu or -1 to exit the program
1

Please input the favorite beverage of person #6: Choose 1, 2, 3, or 4 from the
above menu or -1 to exit the program
-1

The total number of people surveyed is 5. The results are as follows:

Beverage        Number of Votes
********************************
Coffee          3
Tea             0
Coke            1
Orange Juice    1
```

Option 2: Suppose Dave drops a watermelon off a high bridge and lets it fall until it hits the water. If we neglect air resistance, then the distance d in meters fallen by the watermelon after t seconds is d = 0.5 * g * t^2, where the acceleration of gravity g = 9.8 meters/second ^2. Write a program that asks the user to input the number of seconds that the watermelon falls and the height h of the bridge above the water. The program should then calculate the distance fallen for each second from t = 0 until the value of t input by the user. If the total distance fallen is greater than the height of the bridge, then the program should tell the user that the distance fallen is not valid.

Sample Run 1:

```
Please input the time of fall in seconds:
2
Please input the height of the bridge in meters:
100

Time Falling (seconds) Distance Fallen (meters)
*******************************************
0                      0
1                      4.9
2                      19.6
```

Sample Run 2:

```
Please input the time of fall in seconds:
4
Please input the height of the bridge in meters:
50

Time Falling (seconds) Distance Fallen (meters)
*******************************************
0                      0
1                      4.9
2                      19.6
3                      44.1
4                      78.4

Warning-Bad Data: The distance fallen exceeds the height of the bridge
```

Option 3: Write a program that prompts the user for the number of tellers at Nation's Bank in Hyatesville that worked each of the last three years. For each worker the program should ask for the number of days out sick for each of the last three years. The output should provide the number of tellers and the total number of days missed by all the tellers over the last three years.

See the sample output below.

Sample Run:

```
How many tellers worked at Nation's Bank during each of the last three years ?
2

How many days was teller 1 out sick during year 1 ?
5
How many days was teller 1 out sick during year 2 ?
8
How many days was teller 1 out sick during year 3 ?
2

How many days was teller 2 out sick during year 1 ?
1
How many days was teller 2 out sick during year 2 ?
0
How many days was teller 2 out sick during year 3 ?
3

The 2 tellers were out sick for a total of 19 days during the last three years.
```

LESSON SET

6.1 Introduction to Void Functions (Procedures)

PURPOSE

1. To introduce the concept of void functions (procedures)
2. To work with void functions (procedures) that have no parameters
3. To introduce and work with void functions (procedures) that have pass by value and pass by reference parameters

PROCEDURE

1. Students should read the Pre-lab Reading Assignment before coming to lab.
2. Students should complete the Pre-lab Writing Assignment before coming to lab.
3. In the lab, students should complete labs assigned to them by their instructor.

Contents	Pre-requisites	Approximate completion time	Page number	Check when done
Pre-lab Reading Assignment		20 min.	66	
Pre-lab Writing Assignment	Pre-lab reading	10 min.	72	
LESSON 6.1A				
Lab 6.1 Functions with No Parameters	Confident in use of the control structures	15 min.	72	
Lab 6.2 Introduction to Pass by Value	Basic understanding of pass by value.	35 min.	73	
LESSON 6.1B				
Lab 6.3 Introduction to Pass by Reference	Basic understanding of pass by reference.	25 min.	74	
Lab 6.4 Student Generated Code Assignments	Basic understanding of pass by reference and value.	30 min.	76	

PRE-LAB READING ASSIGNMENT

Modules

A key element of structured (well organized and documented) programs is their modularity: the breaking of code into small units. These units, or **modules**, that do not return a value are called **procedures** in most languages and are called **void functions** in C++. Although procedures is the authors' preferred term, this manual uses the word **function** to describe both void functions (discussed in this lesson set) and **value returning functions** (studied in the next lesson set), as this is the terminology used in C++.

The `void main()` section of our program is a function and, up until now, has been the only coded module used in our programs. We also have used pre-defined functions such as `pow` and `sqrt` which are defined in library routines that we "imported" to our program with the `#include <math.h>` directive. We now explore the means of breaking our own code into modules. In fact, the `main` function should contain little more than "calls" to other functions. We can think of the `main` function as a contractor who hires sub-contractors to perform certain duties: plumbers to do the plumbing, electricians to do the electrical work, etc. The contractor is in charge of the order in which these sub-contract jobs are issued.

The `void main()` function consists mostly of calls to functions just like the contractor issuing commands to sub-contractors to come and do their jobs. A computer basically does many simple tasks (modules) that, when combined, produce a set of complex operations. How one determines what those separate tasks should be is one of the skills learned in software engineering, the science of developing quality software. A good computer program consists of several tasks, or units of code, called modules or functions.

In simple programs most functions are called, or invoked, by the `main` function. Calling a function basically means starting the execution of the instructions contained in that module. Sometimes a function may need information "passed" to it to perform designated tasks.

If a function is to find the square root of a number, then it needs that number passed to it by the calling function. Information is passed to or from a function through **parameters**. Parameters are the components of communication between functions. Some functions do very simple tasks such as printing basic output statements to the screen. These may be instructions to the user or just documentation on what the program will do. Such functions are often called parameter-less functions since they do not need anything passed to them by the calling procedure.

Example:

```
#include <iostream.h>

void Description();   //Function prototype

void main()
{
   cout << "Welcome to the Pay Roll Program" << endl;
   Description();     //Call to the function
   cout << "We hoped you enjoyed this program" << endl;

}
```

```
void Description()    //The function heading
{
   cout << "***********************************************" << endl << endl;
   cout << "This program takes two numbers (pay rate & hours)" << endl;
   cout << "and outputs gross pay " << endl;
   cout << "***********************************************" << endl << endl;
}
```

In the above example, three areas have been highlighted. Starting from the bottom we have the function itself which is often called the function definition.

The function **heading** `void Description()` consists of the name of the function preceded by the word `void`. The word `void` means that this function will not return a value to the module that called it.[1] The function name is followed by a set of parentheses. Just like the `main` function, all functions begin with a left brace and end with a right brace. In between these braces are the instructions of the function. In this case they consist solely of `cout` statements that tell what the program does.

Notice that this function comes after the `main` function. How is this function activated? It must be called by either the `main` function or another function in the program. This function is called by `main` with the simple instruction `Description();`.

A **call** to a function could be classified as the sixth fundamental instruction (see Lesson Set 2). Notice that it consists only of the name of the function (not the word `void` preceding it) followed by the set of parentheses and a semicolon. By invoking its name in this way, the function is called at that moment. The program executes the body of instructions found in that function and then returns to the calling function (`main` in this case) where it executes the remaining instructions following the call. Let us examine the order in which the instructions are executed.

The `main` function is invoked which then executes the following instruction:

```
cout << "Welcome to the Pay Roll Program" << endl;
```

Then the call to the function `Description` is encountered which invokes the following instructions:

```
cout << "***********************************************" << endl << endl;
cout << "This program takes two numbers (pay rate & hours)" << endl;
cout << "and outputs gross pay " << endl;
cout << "***********************************************" << endl << endl;
```

After all the instructions in `Description` are executed, control returns to `main` and the next instruction after the call is executed:

```
cout << "We hoped you enjoyed this program" << endl;
```

The first highlighted section of the example is found before `main()` in what we call the global section of the program. It is called a **prototype** and looks just like the function heading except it has a semicolon at the end. Since our example has

[1] In the next lesson set we will see that the word preceding the name of a function can be the data type of the value that the function will return to the calling function.

the "definition of the function" after the call to the function, the program would give us an error when we tried to call it if we did not have some kind of signal to the computer that the definition would be forthcoming. That is the purpose of the prototype. It is a promise (contract if you will) to the compiler that a `void` function called `Description` will be defined after the `main` function. If the `Description` function had been placed in the file before the `main` function which calls it, then the prototype would not have been necessary. However, most C++ programs are written with prototypes so that `main()` can be the first function.

Pass by Value

The following program, Sample Program 6.1, is an extension of the code above. This program will take a pay rate and hours worked and produce gross pay based on those numbers. This can be done in another function called `Paycheck`.

Sample Program 6.1:

```
#include <iostream.h>

//function prototypes
void Description();
void Paycheck(float, int);

void main()
{
   float payrate;
   int hours;

   cout << "Welcome to the Pay Roll Program" << endl;
   Description(); //Call to Description function
   cout << endl << "Please input the pay per hour" << endl;
   cin >> payrate;
   cout << endl << "Please input the number of hours worked" << endl;
   cin >> hours;
   cout << endl << endl;
   Paycheck(payrate,hours); //Call to Paycheck function
   cout << "We hoped you enjoyed this program" << endl;

}

void Description() //The Description function
{
   cout << "************************************************" << endl << endl;
   cout << "This program takes two numbers (pay rate & hours)" << endl;
   cout << "and outputs gross pay "<< endl;
   cout << "************************************************" << endl << endl;

}
```

```
void Paycheck(float rate, int time)
 //This function takes rate and time and multiplies them to get
 //gross pay. It then outputs the result
{
   float gross;
   gross = rate * time;
   cout << "The pay is " << gross << endl;
}
```

The bold sections of this program show the development of another function. This function is a bit different in that it has parameters inside the parentheses of the call, the heading and the prototype. Recall that parameters are the components of communication to and from a function and the call to that function. The function `Paycheck` needs information from the calling routine. In order to find the gross pay it needs the rate per hour and the number of hours worked to be passed to it. The call provides this information by having parameters inside the parentheses of the call `Paycheck(payrate,hours);`. Both `payrate` and `hours` are called **actual parameters**. They match in a one-to-one correspondence with the parameters in the function heading which are called `rate` and `time`:

```
void Paycheck(float rate, int time)
```

The parameters in a function heading are called **formal parameters**.

It is important to compare the call with the function heading.

Call	**Function heading**
`Paycheck(payrate,hours);`	`void Paycheck(float rate, int time)`

1. The call does not have any word preceding the name whereas the function heading has the word `void` preceding its name.
2. The call must NOT give the data type before its actual parameters whereas the heading MUST give the data type of its formal parameters.
3. Although the formal parameters may have the same name as their corresponding actual parameters, they do not have to be the same. The first actual parameter, `payrate`, is paired with `rate`, the first formal parameter. This means that the value of `payrate` is given to `rate`. The second actual parameter, `hours`, is paired with `time` the second formal parameter and gives `time` its value. Corresponding (paired) parameters must have the same data type. Notice that `payrate` is declared as `float` in the `main` function and thus it can legally match `rate` which is also declared as `float` in the function heading. `hours` is declared as `int` so it can be legally matched (paired) with `time` which is declared as `int` in the function heading.
4. The actual parameters (`payrate` and `hours`) pass their values to their corresponding formal parameters. Whatever value is read into `payrate` in the `main` function will be given to `rate` in the `Paycheck` function. This is called **pass by value**. It means that `payrate` and `rate` are two distinct memory locations. Whatever value is in `payrate` at the time of the call will be placed in `rate`'s memory location as its initial value. It should be noted that if function `Paycheck` were to alter the value of `rate`, it would not affect the value of `payrate` back in the `main` function. In essence, pass by value is like making a copy of the value in `payrate` and placing it in `rate`. Whatever is done to that copy in `rate` has no effect on the value in `payrate`. Recall that a formal parameter could have the same name as its

corresponding actual parameter; however, they would still be two different locations in memory.

How does the computer know which location to go to if there are two with the same name? The answer is found in a concept called **scope**. Scope refers to the time and location that a variable or memory location is active in a program. All variables declared in the `main` function become inactive when a function is called and are reactivated when the control returns to `main`. By the same token, all formal parameters and variables declared inside a function are active only during the time the function is executing. What this means is that an actual parameter and its corresponding formal parameter are never active at the same time. Thus there is no confusion as to which memory location to access even if corresponding parameters have the same name. More on scope will be presented in the next lesson set.

It is also important to compare the prototype with the heading.

Prototype	**Function heading**
`void Paycheck(float, int);`	`void Paycheck(float rate, int time)`

1. The prototype has a semicolon at the end and the heading does not.
2. The prototype lists only the data type of the parameters and not their name. However, the prototype could list both and thus be exactly like the heading except for the semicolon. Some instructors tell students to copy the prototype without the semicolon and paste it to form the function heading.

Let us look at all three parts—prototype, call and heading:

1. The heading MUST have both data type and name for all its **formal parameters**.
2. The prototype must have the data type and could have the name for its **formal parameters**.
3. The call MUST have the name but MUST NOT have the data type for its **actual parameters**.

Pass by Reference

Suppose we wanted the `Paycheck` function to only compute the gross pay and then pass this value back to the calling function rather than printing it. We would need another parameter, not to get information from the call but to give information back to the call. This particular parameter could not be **passed by value** since any change made in a function to a *pass by value formal parameter* has no effect on its corresponding actual parameter. Instead, this parameter would be **passed by reference**, which means that the calling function will give the called function the location of its actual parameter instead of a copy of the value that is stored in that location. This then allows the called function to go in and change the value of the actual parameter.

Example: Assume that I have a set of lockers each containing a sheet of paper with a number on it. Making a copy of a sheet from a particular locker and giving that sheet to you will ensure that you will not change my original copy. This is pass by value. On the other hand, if I give you a spare key to a particular locker, you could go to that locker and change the number on the sheet of paper located there. This is pass by reference.

How does the program know whether a parameter is passed by value or by reference? All parameters are passed by value unless they have the character &, which indicates pass by reference, listed after the data type.

Sample Program 6.2:

```
#include <iostream.h>
#include <iomanip.h>

//Function prototypes
void Description();
void Paycheck(float, int, float&);

void main()
{
   float payrate;
   float grosspay;
   float netpay;
   int hours;

   cout << setprecision(2) << setiosflags(ios::fixed);
   cout << "Welcome to the Pay Roll Program" << endl;
   Description(); //Call to Description function
   cout << "Please input the pay per hour" << endl;
   cin >> payrate;
   cout << endl << "Please input the number of hours worked" << endl;
   cin >> hours;
   cout << endl << endl;
   Paycheck(payrate,hours,grosspay);
   netpay = grosspay - (grosspay * .2);
   cout << "The net pay is " << netpay << endl;
   cout << "We hoped you enjoyed this program" << endl;

}

void Description() //The Description function
{
  cout << "************************************************" << endl << endl;
  cout << "This program takes two numbers (payrate and hours)" << endl;
  cout << "and outputs net pay "<< endl;
  cout << "************************************************" << endl << endl;
}

void Paycheck(float rate, int time, float& gross)
 //This function takes rate and time and multiplies them to
 //get gross pay which is returned to the calling function.
{

  gross = rate * time;

}
```

Notice that the function `Paycheck` now has three parameters. The first two, `rate` and `time`, are passed by value while the third has an & after its data type indicating that it is pass by reference. The actual parameter `grosspay` is paired with `gross` since they both are the third parameter in their respective lists. But since this pairing is pass by reference, these two names refer to the SAME memory location. Thus what the function does to its formal parameter `gross` changes the value of `grosspay`. After the `Paycheck` function finds `gross`, control goes back to the `main` function that has this value in `grosspay`. `main` proceeds to find the net pay, by taking 20% off the gross pay, and printing it. Study this latest revision of the program very carefully. One of the lab exercises asks you to alter it.

PRE-LAB WRITING ASSIGNMENT

Fill-in-the-Blank Questions

1. The word ______________ precedes the name of every function prototype and heading that does not return a value back to the calling routine.
2. Pass by ______________ indicates that a copy of the actual parameter is placed in the memory location of its corresponding formal parameter.
3. ______________ parameters are found in the call to a function.
4. A prototype must give the ______________ ______________ of its formal parameters and may give their ______________.
5. A ______________ after a data type in the function heading and in the prototype indicates that the parameter will be passed by reference..
6. Functions that do not return a value are often called ______________ in other programming languages.
7. Pass by ______________ indicates that the location of an actual parameter, rather than just a copy of its value, is passed to the called function.
8. A call must have the ______________ of its actual parameters and must NOT have the ______________ ______________ of those parameters.
9. ______________ refers to the time when a variable is "active."
10. ______________ parameters are found in the function heading.

LESSON 6.1A

LAB 6.1 Functions with No Parameters

Retrieve program `proverb.cpp` from the Lab 6.1 folder. The code is as follows:

```
// This program prints the proverb
// "Now is the time for all good men to come to the aid of their party"
// in a function (procedure) called Proverb that is called by the main function

// PLACE YOUR NAME HERE

#include <iostream.h>

void Proverb();   //This is the prototype for the Proverb function

void main()
{
```

```
    // Fill in the code to  call the  Proverb function

}

// Fill in the code for the function heading and the body of the function that will print
// to the screen the proverb listed in the comments at the beginning of the program
```

Exercise 1: Fill in the blanks so that the program will print out the proverb listed in the comments at the beginning of the program. The proverb will be printed by the function which is called by the `main` function.

LAB 6.2 Introduction to Pass by Value

Retrieve program `newproverb.cpp` from the Lab 6.1 folder. The code is as follows:

```
// This program will allow the user to input from the keyboard
// whether the last word to the following proverb should be party or country:
// "Now is the time for all good men to come to the aid of their _______"
// Inputting a 1 will use the word party. Any other number will use the word country.

// PLACE YOUR NAME HERE

#include <iostream.h>

// Fill in the prototype of the function Proverb.

void main ()
{

    int wordCode;

    cout << "Given the phrase:" << endl;
    cout << "Now is the time for all good men to come to the aid of their ___" << endl;
    cout << "Input a 1 if you want the sentence to be finished with party" << endl;
    cout << "Input any other number for the word country" << endl;
    cout << "Please input your choice now" << endl;
    cin  >> wordCode;
    cout << endl;

    Proverb(wordCode);

}

void Proverb (int number)

// This function takes number from the call and if that number is a 1
// it prints "Now is the time for all good men to come to the aid of their party"
// otherwise it prints "Now is the time for all good men to come to the aid of their
// country"

{
   // Fill in the body of the function to accomplish what is described above
}
```

Exercise 1: Some people know this proverb as "Now is the time for all good men to come to the aid of their country" while others heard it as "Now is the time for all good men to come to the aid of their party." This program will allow the user to choose which way they want it printed. Fill in the blanks of the program to accomplish what is described in the program comments. What happens if you inadvertently enter a float such as -3.97?

Exercise 2: Change the program so that an input of 1 from the user will print "party" at the end, a 2 will print "country" and any other number will be an invalid choice so that the user will need to repeat their choice.

Sample Run:

```
Given the phrase:
Now is the time for all good men to come to the aid of their __
Input a 1 if you want the sentence to be finished with party
Input a 2 if you want the sentence to be finished with country
Please input your choice now
4
I'm sorry but that is an incorrect choice; Please input a 1 or 2
2
Now is the time for all good men to come to the aid of their country
```

LESSON 6.1B

Lab 6.3 Introduction to Pass by Reference

Retrieve program `paycheck.cpp` from the Lab 6.1 folder. This program is similar to Sample Program 6.1 that was given in the Pre-lab Reading Assignment. The code is as follows:

```
// This program takes two numbers (pay rate & hours)
// and multiplies them to get grosspay
// it then calculates net pay by subtracting 15%

// PLACE YOUR NAME HERE

#include <iostream.h>

//Function prototypes
void Description();
void Paycheck(float, int, float&);

void main()
   {
      float payrate;
      float grosspay;
      float netpay;
      int hours;

      cout << "Welcome to the Pay Roll Program" << endl;
      Description(); //Call to Description function
      cout << "Please input the pay per hour" << endl;
```

```
        cin >> payrate;
        cout << endl << "Please input the number of hours worked" << endl;
        cin >> hours;
        cout << endl << endl;
        Paycheck(payrate,hours,grosspay);
        //  Fill in the code to output grosspay

        cout << "The net pay is $" << netpay << endl;
        cout << "We hoped you enjoyed this program" << endl;

    }

void Description() //The function heading
{
    cout << "***********************************************"  << endl << endl;
    cout << "This program takes two numbers (pay rate & hours)" << endl;
    cout << "and  multiplies them to get grosspay "             << endl;
    cout << "it then calculates net pay by subtracting 15%"     << endl;
    cout << "***********************************************"  << endl << endl;
}

void Paycheck(float rate, int time, float& gross)

 // This function takes rate and time and multiplies them to
 // get gross pay and then finds net pay by subtracting 15%
 // Both values are returned to the calling function.
{
    //  Fill in the code to find grosspay and net pay

}
```

Exercise 1: Fill in the code (places in bold) and change the program so that the function `Paycheck` has a fourth parameter called `net` that determines the net pay by subtracting 15 percent from the gross pay. Both `gross` and `net` should be returned to the `main()` function where those values are printed.

Exercise 2: Compile and run your program with the following data and make sure you get the output shown.

```
Please input the pay per hour
9.50
Please input the number of hours worked
40

The gross pay is $380
The net pay is $323
We hoped you enjoyed this program
```

Exercise 3: Are the parameters `gross` and `net` in the modified `Paycheck` function you created in Exercise 1 above, pass by value or pass by reference?

____

Exercise 4: Alter the program so that `gross` and `net` are printed in the function `Paycheck` instead of in `main()`. The `main()` function executes only the statement

```
cout << "We hoped you enjoyed this program" << endl;
```

after the return from the function `Paycheck`.

Exercise 5: Run the program again using the data from Exercise 2. You should get the same results.

LAB 6.4 Student Generated Code Assignments

Option 1: Write a program that will read two floating point numbers (the first read into a variable called `first` and the second read into a variable called `second`) then calls the function `swap` with the actual parameters `first` and `second`. The `swap` function having formal parameters `number1` and `number2` should swap the value of the two variables. Note: This is similar to a program you did in Lesson Set 1; however, now you are required to use a function. You may want to look at `logicprob.cpp` from Lesson Set 1.

Sample Run:

```
Enter the first number
Then hit enter
80
Enter the second number
Then hit enter
70

You input the numbers as 80 and 70.
After swapping, the first number has the value of 70 which was the value of the
second number.
The second number has the value of 80 which was the value of the first number.
```

Exercise 1: Compile the program and correct it if necessary until you get no syntax errors.

Exercise 2: Run the program with the sample data above and see if you get the same results.

Exercise 3: The `swap` parameters must be passed by ________________. Why?

Option 2: Write a program that will input miles traveled and hours spent in travel. The program will determine miles per hour. This calculation must be done in a function other than `main`; however, `main` will print the calculation. The function will thus have 3 parameters: `miles`, `hours`, `milesPerHour`. Which parameter(s) are pass by value and which are (is) passed by reference? Output is fixed with 2 decimal point precision.

Sample Run:

```
Please input the miles traveled
475
Please input the hours traveled
8
Your speed is 59.38 miles per hour
```

Option 3: Write a program that will input grades, the number of which is also input by the user. The program will find the sum of those grades and pass it, along with the number of grades, to a function that has a "pass by reference" parameter that will contain the numeric average of those grades as processed by the function. The `main` function will then determine the letter grade of that average based on a 10-point spread.

90–100	A
80–89	B
70–79	C
60–69	D
0–59	F

Sample Run:

```
Enter the number of grades
3
Enter a numeric grade between 0-100
90
Enter a numeric grade between 0-100
80
Enter a numeric grade between 0-100
50
The grade is C
```

LESSON SET

6.2 Functions that Return a Value

PURPOSE

1. To introduce the concept of scope
2. To understand the difference between static, local and global variables
3. To introduce the concept of functions that return a value
4. To introduce the concept of overloading functions

PROCEDURE

1. Students should read the Pre-lab Reading Assignment before coming to lab.
2. Students should complete the Pre-lab Writing Assignment before coming to lab.
3. In the lab, students should complete labs assigned to them by the instructor.

Contents	Pre-requisites	Approximate completion time	Page number	Check when done
Pre-lab Reading Assignment		20 min.	80	
Pre-lab Writing Assignment	Pre-lab reading	10 min.	87	
LESSON 6.2A				
Lab 6.5 Scope of Variables	Basic understanding of scope rules	15 min.	87	
Lab 6.6 Parameters and Local Variables	Basic understanding of formal and actual parameters and local variables	35 min.	89	
LESSON 6.2B				
Lab 6.7 Value Returning and Overloading Functions	Understanding of value returning functions Understanding of overloading functions	30 min.	91	
Lab 6.8 Student Generated Code Assignments	Basic understanding of pass by reference and value.	30 min.	94	

PRE-LAB READING ASSIGNMENT

Scope

As mentioned in Lesson Set 6.1, the scope of an identifier (variable, constant, function, etc.) is an indication of where it can be accessed in a program. There can be certain portions of a program where a variable or other identifier can not be accessed for use. Such areas are considered out of the scope for that particular identifier. The header (the portion of the program before `main`) has often been referred to as the global section. Any identifier defined or declared in this area is said to have **global scope**, meaning it can be accessed at any time during the execution of the program. Any identifier declared outside the bounds of all the functions have global scope. Although most constants and all functions are declared globally, variables should almost **never** be declared in this manner. **Local scope** refers to identifiers defined within a **block**. They are active only within the bounds of that particular block. In C++ a block begins with a left brace { and ends with a right brace }. Recall that all functions (including `main`) must begin and end with a pair of braces which form a block. Other blocks could be defined within blocks and thus the scope of some identifiers would not extend for the entire function.

Sample Program 6.2a

```
#include <iostream.h>

const PI = 3.14;

void Heading();

void main()
{
   float circle;
   cout << "circle has local scope that extends the entire main function"
        << endl;
   {
     float square;
     cout << "Square has local scope active for only a portion of main" << endl;
     cout << "Both square and circle can be accessed here "
          << "as well as the Global constant PI" << endl;
   }

   cout << "circle is active here, but square is not" << endl;
   Heading();

}

void Heading()
{
    int triangle;

    cout << "The Global constant PI is active here"
         << " as well as the local variable triangle" << endl;
}
```

Notice that the nested braces within the outer braces of `main()` indicate another block in which `square` is declared. `square` is active only within the bounds of the inner braces while `circle` is active for the entire `main` function. Neither of these are active when the function `Heading` is called. `triangle` is a local variable of the function `Heading` and is active only when that function is active. `PI`, being a global identifier, is active everywhere.

Formal parameters (Lesson Set 6.1) have the same scope as local variables declared in the outmost block of the function. That means that the scope of formal parameters of a function is the entire function. The question may arise about variables with the same name. For example, could a local variable in the function `Heading` of the above example have the name `circle`? The answer is yes, but it would be a different memory location than the one declared in the `main` function. There are rules of **name precedence** which determine which memory location is active among a group of 2 or more variables with the same name. The most recently declared variable has precedence over any other variable with the same name. In the above example, if `circle` had been declared in the `Heading` function then the memory location assigned with that declaration would take precedence over the location declared in `main()` as long as the function `Heading` was active.

Lifetime is another way of describing the scope of a variable. It refers to the time during a program that an identifier has storage assigned to it.

Scope Rules

1. The scope of a global identifier, any identifier declared outside all functions, is the entire program.
2. Functions are declared globally. That means any function can call any other function at any time.
3. The scope of a local identifier is from the point of its declaration to the end of the block in which it is declared. This includes any nested blocks that may be contained within, unless the nested block has a variable declared in it that has the same name.
4. Scope of formal parameters is the same as the scope of local variables of the function.

Why are variables almost never declared globally? Good structured programming assures that all communication between functions will be explicit through the use of parameters. Global variables can be changed by any function. In large projects, where more than one programmer may be working on the same program, global variables are unreliable since their values can be changed by any function or any programmer and are thus not used. The inadvertent changing of global variables in a particular function can cause unwanted side effects.

Static Local Variables

One of the biggest advantages of a function is the fact that it can be called multiple times to perform its job. This saves much needed time and memory space. The value of local variables do not remain between multiple function calls. What this means is that the value assigned to a local variable of a function is lost once the function is finished executing. If the same function is called again that value will not necessarily be present for that local variable. Local variables start "fresh," in terms of their value, each time the function is called. There may be times when a function needs to retain the value of a variable between calls. This can be done by declaring it to be a **static variable**. Static variables are declared with the word **`static`** preceding the data type and name of the variable. Thus the lifetime of a static variable is different than a local variable. A static variable's memory space is retained after the function in which it is declared has finished executing.

Default Arguments

Actual parameters (parameters used in the call to a function) are often called **arguments**. Normally the number of actual arguments must equal the number of formal parameters, and it is good programming practice to use this one-to-one correspondence between actual and formal parameters. It is possible, however, to assign default values to all formal parameters so that the calling instruction does not have to pass values for all the arguments. Although these default values can be expressed in the function heading, they are usually defined in the prototype. Certain arguments can be left out; however, if an argument is left out, then all the following arguments must also be left out. For this reason, pass by reference arguments should be placed first (since by their very nature they must be included in the call).

Sample Program 6.2b:

```
#include <iostream.h>
#include <iomanip.h>

void Netpay(float& net,  int hours = 40,  float rate = 6.00);

void main()

{
   int hoursworked = 20;
   float payrate = 5.00;
   float paynet;

   cout << setprecision(2) << setiosflags(ios::fixed);

   Netpay(paynet);           // note the call only involves 1 parameter
   cout << " The net pay is $" << paynet << endl;

}

void Netpay (float& net, int hours, float rate)

{
   net = hours * rate;
}
```

What would happen if `paynet` was also not listed in the calling instruction? An error would occur that states that the function can not take 0 arguments. The reason for this is that the `net` formal parameter does not have a default value and so the call must have at least one argument. In general there must be as many actual arguments as formal parameters that do not have default values. Of course some or all default values can be overridden.

The following calls would all be legal in the example program. Fill in the values that the `Netpay` function receives for `hours` and `rate` in each case. Also fill in the value that you expect net pay to have for each call.

```
Netpay(paynet);                                  The net pay is $________
   Netpay receives the value of _____ for hours and _____ for rate.

Netpay(paynet,hoursworked);                        The net pay is $________
   Netpay receives the value of _____ for hours and _____ for rate.

Netpay(paynet, hoursworked, payrate);               The net pay is $________
   Netpay receives the value of _____ for hours and _____ for rate.
```

The following would not be correct. List what you think causes the error in each case.

```
Netpay(paynet, payrate);
Netpay(hoursworked, payrate);
Netpay(payrate);
Netpay();
```

Functions that Return a Value

The functions discussed in the previous lesson set are not "true functions" because they do not return a value to the calling function. They are often referred to as procedures in computer science jargon. True functions, or value returning functions, are modules that return exactly one value to the calling routine.

```
// Prototype of a user defined function that will cube the value passed to it

int Cubeit(int x);
....
void main()
{
   int x = 2;
   int cube;

   cube = Cubeit(x);    // The call to Cubeit function
   ......
}
int Cubeit(int x) // Notice the function type is int rather than void.
{
   int num;

   num = x * x * x;
   return num;
}
```

The function `Cubeit` receives the value of `x`, which in this case is 2, and finds its cube which it places in a local variable `num`. It then returns the value stored in `num` to the call `Cubeit(x)`. The value 8 replaces the entire call and is assigned to `cube`, i.e., `cube = Cubeit(x)` is replaced with `cube = 8`.

Value returning functions replace the word `void` with the data type of the value that is returned. Since these functions return one value, there should be no effect on any parameters that are passed to it from the call. What this means is that all parameters of this type of function should be pass by value, NOT pass by reference. Nothing in C++ prevents the programmer from using pass by reference in value returning functions; however, they should never be used.

The `netpay` program (Sample Program 6.2b) has a module that calculates the net pay when given hours worked and the hourly pay rate. Since it calculates only one value that is needed by the call, it could easily be implemented as a value returning function, instead of by having `paynet` passed by reference.

Sample Program 6.2c:

```
#include <iostream.h>
#include <iomanip.h>

float Netpay(int hours,  float rate);

void main()
{
  int hoursworked = 20;
  float payrate = 5.00;
  float paynet;

  cout << setprecision(2) << setiosflags(ios::fixed);

  paynet = Netpay(hoursworked,payrate);
  cout << " The net pay is $" << paynet << endl;

}

float Netpay(int hours,  float rate)
{
   float netpay;
   netpay = hours * rate;
   return netpay;
}
```

Notice how this function is called, i.e., `paynet = Netpay(hoursworked,payrate);`. The call to the function is not a stand-alone statement, but rather part of an assignment statement. The call is used in an expression. In fact, the function will return a floating value (`netpay`) that replaces the entire right-hand side of the assignment statement. This is the first major difference between the two types of functions (`void` functions and value returning functions). A `void` function is called by just listing the name of the function along with its arguments. A value returning function is called within a portion of some fundamental instruction (the right-hand side of an assignment statement, condition of a selection or loop statement, or argument of a `cout` statement). As mentioned earlier, another difference is that in both the prototype and function heading the word `void` is replaced with the data type of the value that is returned. A third difference is the fact that a value returning function MUST have a `return` statement. It is usually the very last instruction of the function. The function `Netpay` could have had just the return instruction, `return hours * rate;`. This would have returned the value found by multiplying `hours` and `rate` to the calling instruction. The following is a comparison between the implementation as a procedure (`void` function) and as a value returning function.

	Value Returning Function	**Procedure**
PROTOTYPE	float Netpay(int hours, float rate);	void Netpay(float& netpay, int hours, float rate);
CALL	netpay=Netpay(hoursworked, rate);	Netpay(netpay, hoursworked, rate);
HEADING	float Netpay(int hours, float rate)	void Netpay(float& netpay, int hours, float rate)
BODY	{ return hours * rate; }	{ netpay = hours * rate; }

Functions can also return a Boolean data type to test whether a certain condition exists (true) or not (false).

Overloading Functions

Uniqueness of identifier names is a vital concept in programming languages. The convention in C++ is that every variable, function, constant, etc. name needs to be unique. However, there is an exception. Two or more functions may have the same name as long as their parameters differ in quantity or data type. For example, a programmer could have two functions with the same name that do the exact same thing on different data types.

Example: Look at the following prototypes of functions all having the same name and yet all could be included in the same program because each one differs from all the others either by the number of parameters or the data types of the parameters.

```
int   Add(int a, int b, int c);
int   Add(int a, int b);
float Add(float a, float b, float c);
float Add(float a, float b);
```

Stubs and Drivers

Many IDEs (Integrated Development Environments) have software debuggers which are used to help in the location of logical errors; however, programmers often use the concept of stubs and drivers to test and debug programs that use functions and procedures. A **stub** is nothing more than a dummy function that is called instead of the actual function. It usually does little more than write a message to the screen indicating that it was called with certain arguments. In structured design, the programmer often wants to delay the implementation of certain details until the overall design of the program is complete. The use of stubs makes this possible.

Sample Program 6.2d:

```
#include <iostream.h>

int Squarert(int x);    //A user defined prototype of a function that will find
                        //the square root of a number

void main()
```

continues

```
{
   int number1;
   int squarednum;

   cout << "Please input the number you would like the square root of" << endl;
   cout << "Input a -99 when you would like to quit" << endl;
   cin >> number1;
   while (number1 != 99)
   {
     cout << endl << "The square root of your number is "
     << Squarert(number1) << end;
     cout << "Please input the number you would like the square root of" << endl;
     cout << "Input a -99 when you would like to quit" << endl;
     cin >> number1;
   }

}

int Squarert(int x)
{
   cout << "Squarert function was called with " << x << " as its argument";

} //This bold section is the stub.
```

This example shows that the programmer can test the execution of `main` and call to the function without having the function finding the square root. This allows the programmer to concentrate on one component at a time. Although a stub is not really needed in this simple program, stubs are very useful for larger programs.

A **driver** is a module that tests a function by simply calling it. While one programmer may be working on the `main` function, another programmer could be developing the code for a particular function. In this case the programmer would not be so concerned with the calling of the function but rather with the body of the function itself. In such a case a driver (call to the function) could be used just to see if the function performs properly.

Sample Program 6.2e:

```
#include <iostream.h>
#include <math.h>

int Squarert(int x);    //A user defined prototype of a function that will find
                        //the square root of a number

void main()
{
   int number1;

   cout << "Calling Squarert function with a 4" << endl;
   cout << "This is the result " << Squarert(4) << endl;
```

```
    .....

}

int Squarert(int x)
{
   return sqrt(x);
}
```

In this example, the `main` function is used solely as a tool (driver) to call the `Squarert` function to see if it performs properly.

PRE-LAB WRITING ASSIGNMENT

Fill-in-the-Blank Questions

1. Variables of a function that retain their value over multiple calls to the function are called _______________ variables.
2. In C++ all functions have _______________ scope.
3. Default arguments are usually defined in the _______________ of the function.
4. A function returning a value should never use pass by _______________ parameters.
5. Every function that begins with a data type in the heading, rather than the word `void`, must have a(n) _______________ statement somewhere, usually at the end, in its body of instructions.
6. A _______________ is a program that tests a function by simply calling it.
7. In C++ a block boundary is defined within a pair of _______________.
8. A _______________ is a dummy function that just indicates that a function was called properly.
9. Default values are generally not given for pass by _______________ parameters.
10. _______________ functions are functions that have the same name but a different parameter list.

LESSON 6.2A

LAB 6.5 Scope of Variables

Retrieve program `scope.cpp` from the Lab 6.2 folder. The code is as follows:

```
// This program will demonstrate the scope rules.

// PLACE YOUR NAME HERE

#include <iostream.h>
#include <iomanip.h>

const double PI = 3.14;
const double RATE = 0.25;
```

continues

```
void Area(float, float&);
void Circumference(float, float&);

void main()

{

   cout << setiosflags(ios::fixed) << setiosflags(ios::showpoint) << setprecision(2);

   float radius = 12;

   cout <<" Main function Outer layer" << endl;
   cout <<" LIST THE IDENTIFIERS THAT are active here" << endl << endl;
   {
      float area;

      cout << "Main function first inner block" << endl;
      cout << "LIST THE IDENTIFIERS THAT are active here" << endl << endl;

      // Fill in the code to call Area here

      cout << "The radius = " << radius << endl;
      cout << "The area = " << area << endl << endl;
   }

   {
      float radius = 10;
      float circumference;

      cout << "Main function second inner block" << endl;
      cout << "LIST THE IDENTIFIERS THAT are active here" << endl << endl;

      // Fill in the code to call Circumference here

      cout << "The radius = " << radius << endl;
      cout << "The circumference = " << circumference << endl << endl;

   }

      cout << "Main function after all the calls" << endl;
      cout << "LIST THE IDENTIFIERS THAT are active here" << endl << endl;

}

void Area(float rad, float& answer)
{

   cout << "AREA FUNCTION" << endl << endl;
   cout << "LIST THE IDENTIFIERS THAT are active here" << endl << endl;
   // FILL in the code given the rad as radius that will find the area
   // to be stored in answer
```

```
}

void Circumference(float length, float& distance)

{
    cout << "CIRCUMFERENCE FUNCTION" << endl << endl;
    cout << "LIST THE IDENTIFIERS THAT are active here" << endl << endl;
    // FILL in the code given the length as radius that will find the
    // circumference to be stored in distance

}
```

Exercise 1: Fill in the following chart by listing all the identifiers (function names, variables, constants)

GLOBAL	Main	Main(inner 1)	Main (inner 2)	Area	Circumference

Exercise 2: For each `cout` instruction that reads:

```
cout << " LIST THE IDENTIFIERS THAT are active here" << endl;
```

Replace the words in all caps by a list of all identifiers active at that location. Change it to have the following form:

```
cout << "area, radius and PI are active here" << endl;
```

Exercise 3: For each comment in bold, place the proper code to do what it says.

NOTE: area = πr^2
circumference = $2\pi r$

Exercise 4: Before compiling and running the program, write out what you expect the output from this program to be.

What value for `radius` will be passed by `main` (first inner block) to the Area function?

What value for `radius` will be passed by `main` function (second inner block) to the `Circumference` function?

Exercise 5: Compile and run your program. Your instructor may ask to see your program run or obtain a hard copy of your work.

LAB 6.6 Parameters and Local Variables

Retrieve program `money.cpp` from the Lab 6.2 folder. The code is as follows:

```
// PLACE YOUR NAME HERE

#include <iostream.h>
#include <iomanip.h>

void Normalizemoney(float& dollars, float& total,  int cents = 1089 );

void main()

{
    int cents;
    float dollars;
    float total;

    cout << setiosflags(ios::fixed) << setiosflags(ios::showpoint) << setprecision(2);

    cents = 95;
    Normalizemoney( dollars, total,cents);

    cout << "Our total so far is  $" << dollars << endl;

    cents = 193;
    Normalizemoney(dollars, total,cents);

    cout << "We have added another $" << dollars <<"  to our total" << endl;
    cout << "Our total so far is  $" << total << endl;

    Normalizemoney(dollars,total);  //The default value used for 3rd paramter

    cout << "We have added another $" << dollars <<"  to our total" << endl;
    cout << "Our total is  $" << total << endl;

}

void Normalizemoney(float& dollars, float& total,  int cents )

{

    // This function is given cents
    // It will convert cents to dollars and send this to the calling function
    // It will keep a running total of all the money it processed in total

    // Sum is a local static variable that keeps a running total of all money
    // processed and sets its value to total

    // Fill in declaration of sum
    _________________ sum = 0.0;

    // Fill in the code to convert cents to dollars
```

```
    sum = sum + dollars; // This adds dollars to the sum
    total = sum;         // Total gets the value of sum to be sent back to main

}
```

Exercise 1: You will notice that the function has to be completed. This function will take `cents` and convert it to `dollars`. It also keeps a running total of all the money it has processed. Assuming that the function is complete, write out what you expect the program will print.

Exercise 2: Complete the function. Fill in the blank space to declare `sum` and then write the code to convert `cents` to `dollars`. Example: 789 cents would convert to 7.89. Compile and run the program to get the expected results. Think about how `sum` should be declared.

LESSON 6.2B

LAB 6.7 Value Returning and Overloading Functions

Retrieve program `convertmoney.cpp` from the Lab 6.2 folder. The code is as follows:

```
// This program will input American money and convert it to foreign currency

// PLACE YOUR NAME HERE

#include <iostream.h>

//Prototypes of the functions
void  MultiConversion(float dollars, float& lire, float& pesos);
void  MultiConversion(float dollars, float& lire, float& pesos, float& francs);
float FrancsConversion(float dollars);
float LireConversion(float dollars);
float PesosConversion(float dollars);

void main ()

{
    float dollars;
    float lire;
    float pesos;
    float francs;

    cout << "Please input the amount of American Dollars you want converted"
         << " to lire and pesos" << endl;
    cin >> dollars;

    // Drivers!!!!  Replace these with actual calls to and output from the functions once
    // they are created

    cout << "Call to MultiConversion with dollars, lire and pesos" << endl << endl;
    MultiConversion(dollars, lire, pesos);
```

continues

```
    cout << "Please input the amount of American Dollars you want converted"
         << " to lire, pesos and francs" << endl;
    cin >> dollars;

    cout << "Call to MultiConversion with dollars, lire, pesos and francs"
         << endl << endl;
    MultiConversion(dollars, lire, pesos, francs);

    cout << "Please input the amount of American Dollars you want converted"
         << " to francs" <<endl;
    cin >> dollars;

    cout << "Call to FrancsConversion " << FrancsConversion(dollars) << endl << endl;

    cout << "Please input the amount of American Dollars you want converted"
         << " to lire" << endl;
    cin  >> dollars;

    cout << "Call to LireConversion " << LireConversion(dollars) << endl << endl;

    cout << "Please input the amount of American Dollars you want converted"
         << " to pesos " << endl;
    cin >> dollars;

    cout << "Call to  PesosConversion " << PesosConversion(dollars) << endl << endl;

}

//All of the functions are stubs that just serve to test the functions
//Replace with code that will cause the functions to execute properly

void  MultiConversion(float dollars, float& lire, float& pesos)

{
    cout << "The function MultiConversion with dollars,lire and pesos "
         << endl <<" was called with " << dollars <<" dollars" << endl << endl;

}

void  MultiConversion(float dollars, float& lire, float& pesos, float& francs)

{
    cout << "The function MultiConversion with dollars,lire, pesos and francs"
         << endl << " was called with " << dollars << " dollars" << endl << endl;

}

float FrancsConversion(float dollars)

{
    cout << "The function FrancsConversion was called with " << dollars <<" dollars"
         << endl << endl;

    return 0;
}
```

```
float LireConversion(float dollars)
{
   cout << "The function LireConversion was called with " << dollars
        << " dollars" << endl << endl;
   return 0;
}

float PesosConversion(float dollars)

{
   cout << "The function PesosConversion was called with " << dollars
        << " dollars"<<endl;

    return 0;
}
```

Exercise 1: Run this program and see the results. You can input anything that you like for the dollars to be converted. Notice that it has drivers and stubs as well as overloaded and value returning functions. Study the stubs and drivers carefully. Notice that in this case the value returning functions always return 0.

Exercise 2: Complete the program by turning all the stubs into workable functions and the drivers to practical calls to the functions. Be sure to call true functions differently than procedures. Make sure that functions return the converted dollars into the proper currency. Although the exchange rates vary from day to day, use the following conversion chart for the program. These values should be declared as constants in the global section so that any change in the exchange rate can be made there and nowhere else in the program.

One Dollar = 2048 lire
9.815 pesos
6.937 francs

Sample Run:

```
Please input the amount of American Dollars you want converted to lire and pesos
9.35

$9.35 is converted to 19148.8 lire and 91.7703 pesos.

Please input the amount of American Dollars you want converted to lire and pesos
and francs.
10.67
$10.67 is converted to 21852.2 lire, 104.726 pesos and 74.0178 francs

Please input the amount of American Dollars you want converted to francs
12.78
$12.78 is converted to 88.6549 francs

Please input the amount of American Dollars you want converted to lire
2.45
$2.45 is converted to 5017.6 lire

Please input the amount of American Dollars you want converted to pesos
8.75
$8.75 is converted to 85.8812 pesos
```

LAB 6.8 Student Generated Code Assignments

Option 1: Write a program that will convert miles to kilometers and kilometers to miles. The user will indicate both a number (representing a distance) and a choice of whether that number is in miles to be converted to kilometers or kilometers to be converted to miles. Each conversion is done in value returning functions. You can use the following conversions.

1 kilometer = .621 miles
1 mile = 1.61 kilometers

Sample Run:

```
Please input
1 Convert miles to kilometers
2 Convert kilometers to miles
3 Quit

1
Please input the miles to be converted
120
120 miles = 193.2 kilometers.

Please input
1 Convert miles to kilometers
2 Convert kilometers to miles
3 Quit
2
Please input the kilometers to be converted
235
235 kilometers = 145.935 miles.

Please input
1 Convert miles to kilometers
2 Convert kilometers to miles
3 Quit
3
```

Option 2: Write a program that will input the number of wins and losses that a baseball team acquired during a complete **season**. The wins should be input in a parameter-less value returning function that returns the wins to the `main` function. A similar function should do the same thing for the losses. A third value returning function should calculate the percentage of wins. It should receive the wins and losses as parameters and returns the percentage (float) to the `main` program which then prints the result. The percentage should be printed as a percent to two decimal places.

Sample Run:

```
Please input the number of wins
80
Please input the number of losses
40
The percentage of wins is 66.67%
```

Option 3: Write a program that outputs a dentist bill. For members of a dental plan, the bill consists of the service charge (for the particular procedure performed), and test fees, input to the program by the user. To non-members the charges consist of the above services plus medicine (also input by the user). The program first asks if the patient is a member of the dental plan. The program uses two overloaded functions to calculate the total bill. Both are value returning functions that return the total charge.

Sample Run 1:

```
Please input a one if you are a member of the dental plan
Input any other number if you are not
1
Please input the service charge
7.89
Please input the test charges
89.56
The total bill is $97.45
```

Sample Run 2:

```
Please input a one if you are a member of the dental plan
Input any other number if you are not
2
Please input the service charge
75.84
Please input the test charges
49.78
Please input the medicine charges
40.22
The total bill is $165.84
```

LESSON SET

7 Arrays

PURPOSE

1. To introduce and allow students to work with arrays
2. To introduce the `typedef` statement
3. To work with and manipulate multidimensional arrays

PROCEDURE

1. Students should read the Pre-lab Reading Assignment before coming to lab.
2. Students should complete the Pre-lab Writing Assignment before coming to lab.
3. In the lab, students should complete labs assigned to them by the instructor.

Contents	Pre-requisites	Approximate completion time	Page number	Check when done
Pre-lab Reading Assignment		20 min.	98	
Pre-lab Writing Assignment	Pre-lab reading	10 min.	103	
LESSON 7A				
Lab 7.1 Working with One-Dimensional Arrays	Basic understanding of one-dimensional arrays	30 min.	104	
Lab 7.2 Strings as Arrays of Characters	Basic understanding of arrays of characters	20 min.	106	
LESSON 7B				
Lab 7.3 Working with Two-Dimensional Arrays	Understanding of multi-dimensional arrays	40 min.	109	
Lab 7.4 Student Generated Code Assignments	Basic understanding of arrays	30 min.	113	

PRE-LAB READING ASSIGNMENT

One-Dimensional Arrays

So far we have talked about a variable as a single location in the computer's memory. It is possible to have a collection of memory locations, all of which have the same data type, grouped together under one name. Such a collection is called an **array**. Like every variable, an array must be declared so that the computer can "reserve" the appropriate amount of memory. This amount is based upon the type of data to be stored and the number of locations, i.e., size of the array, each of which is given in the declaration.

Example: Given a list of ages (from a file or input from the keyboard), find the number of people for each age.

The programmer would not know the ages to be read but would need a space for the total number of occurrences of each "legitimate age." Assuming that ages 1, 2, . . . , 100 are possible, the following array declaration could be used.

```
const int TOTALYEARS = 100;

void main()
{
   int agefrequency[TOTALYEARS];      //reserves memory for 100 ints

}
```

Following the rules of variable declaration, the data type (integer in this case) is given first, followed by the name of the array (`agefrequency`), and then the total number of memory locations enclosed in brackets. The number of memory locations must be an integer expression greater than zero and can be given either as a named constant (as shown in the above example) or as a literal constant (an actual number such as 100).

Each element of an array, consisting of a particular memory location within the group, is accessed by giving the name of the array and a position with the array (subscript). In C++ the subscript is enclosed in square brackets and the numbering of the subscripts always begins at 0 and ends with one less than the total number of locations.

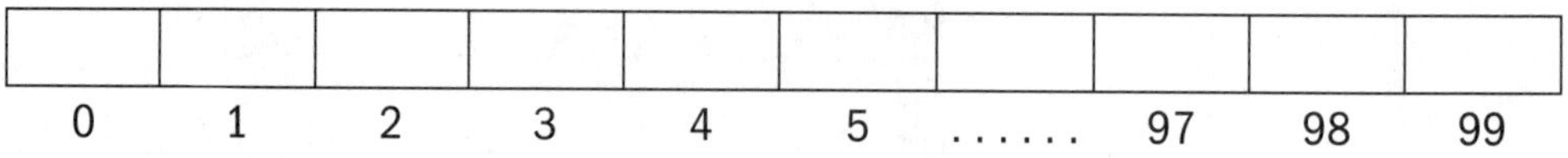

If in our example we wanted ages from 1 to 100, the number of occurrences of age 4 would actually be placed in subscript 3 since it is the "fourth" location in the array. This odd way of numbering is often confusing to new programmers; however, it quickly becomes routine.[1]

[1] Some students actually add one more location and then ignore location 0, letting 1 be the first location. In the above example such a process would use the following declaration: `int agefrequency[101];` and use only the subscripts 1 through 100. Our examples will use location 0. Your teacher will tell you which method to use.

Array Initialization

In our example, we could let `agefrequency[0]` keep a count of how many 1's we read in, `agefrequencey[1]` keep a count of how many 2's we read in, etc. Thus in our example, keeping track of how many people of a particular age exist in the data read in would require reading each age and then adding one to the location holding the count for that age. Of course it is important that all the counters start at 0. The following shows the initialization of all the elements of our sample array to 0.

```
int pos = 0;            //pos serves as the subscript to the array

for (pos = 0; pos < TOTALYEARS; pos++)

   agefrequency[pos] = 0;
```

A simple `for` loop will process the entire array, adding one to the subscript each time through the loop. Notice that the subscript (`pos`) starts with 0. Why is the condition `pos < TOTALYEARS` used instead of `pos <= TOTALYEARS`? Remember that the last subscript is one less than the total number. The subscripts go from 0 to 99 in this case.

Array Processing

Arrays are generally processed inside loops so that the input/output processing of each element of the array can be performed with minimal statements. Our age frequency problem first needs to read in the ages from a file or from the keyboard. For each age read in, the "appropriate" element of the array (the one corresponding to that age) needs to be incremented by one. The following examples show how this can be accomplished:

from a file using `infile` as a logical name

```
pos = 0;




infile >> currentage;
while (infile)

{
  agefrequency[currentage-1] =
  agefrequency[currentage-1] + 1;
  infile >> currentage;





}
```

from a keyboard with –99 as sentinel data

```
pos = 0;
cout << "Please input an age from one"
     << "to 100. Put -99 to stop"
     << endl;
cin >> currentage;

while (currentage != -99)
{
  agefrequency[currentage-1] =
  agefrequency[currentage-1] + 1;
  cout << "Please input an age from "
       <<"one to 100. Put -99 to stop"
       << endl;
cin >> currentage;

}
```

The `while(infile)` statement means that while there is more data in the file `infile`, the loop will continue to process.

Notice that we initialize `pos` (which is the index, or subscript, for our array) to 0. In reading from a file or from the keyboard we **prime the read**,[2] which

[2] Priming the read for a `while` loop means having an input just before the loop condition (just before the `while`) and having another one as the last statement in the loop.

means we read a value before we use our test condition to see if the loop should be executed. When we read an age, we increment the location in the array that keeps track of the amount of people in that age group. Since C++ arrays always start with 0, that location will be at the subscript one value less than the age we read in.

4	0	14	5	0	6		1	0
0 1 year	1 2 years	2 3 years	3 4 years	4 5 years	5 6 years	......	98 99 years	99 100 years

Each element of the array contains the number of people of the given age. The data shown here is from a random sample run. In writing the information stored in the array, we want to make sure that only those array elements that have values greater than 0 are output.

Look at the following code:

```
for (int agecounter = 0; agecounter < TOTALYEARS; agecounter++)

        if (agefrequency[agecounter] > 0)
          cout << "The number of people " << agecounter + 1 <<" years old is "
               << agefrequency[agecounter] << endl;
```

The `for` loop goes from 0 to one less than `TOTALYEARS` (0 to 99). This will test every element of the array. If that element has a value greater than 0, it will be output. What does outputting `agecounter + 1` do? It gives the age we are dealing with at any given time, while the value of `agefrequency[agecounter]` gives the number of people in that age group.

The complete age frequency program will be given as one of the optional lab assignments in Lab 7.4.

Arrays as Arguments

Arrays can be passed as arguments (parameters) to functions. Although variables can be passed by value or reference, arrays are ALWAYS passed by reference, since it is not productive to make a "copy" of all elements of the array. This means that arrays can be altered by the calling function. Although arrays are always passed by reference, they never have the & symbol between the data type and name as other pass by reference parameters have.

Example: Suppose we have an array of grades that we want to pass to a function or procedure to sum all its values.

```
const int TOTALGRADES = 50; //TOTALGRADES is the maximum size of the array

void Sumarray(int size, int grades[], int&  sum); //function prototype

void main()
{
   int grades[TOTALGRADES];
   int numberOfRecs = 0, sum;          // numberOfRecs holds amount of grades
   int indgrade;                       // indgrade holds a grade

   cout << "Please input a grade: type -99 to stop" << endl;
   cin >> indgrade;
```

```
   while ( indgrade != -99)
   {
      grades[numberOfRecs] = indgrade;
      numberOfRecs++;                   // increment subscript of array
      cout << "Please input a grade: type -99 to stop" << endl;
      cin >> indgrade;
   }
}
```

Upon exiting this loop which reads values into an array, the variable used as a subscript to the array (`numOfRecs` in this case) contains the size of the array actually used.

```
// Call to the Function

   Sumarray(numberOfRecs, grades, sum);

   cout << "The sum of all grades is" << sum;

}
```

The call gives only the name of the parameters and not their data types. There is no need to indicate by brackets that `grades` is an array. It is declared as such by the calling module.

```
// Procedure that sums the value of an array:

void Sumarray(int size, int grades[], int&  sum)

{
     sum = 0;
     for (int pos = 0; pos < size; pos++)

         sum = sum + grades[pos];
}
```

The variable `size` contains the number of elements in the array to be processed. In most cases not every element of the array is used, which means the size of the array given in its declaration and the number of actual elements used are rarely the same. For that reason we often pass the actual number of elements used in the array as a parameter to a procedure using that array. The second parameter is the array of grades itself. Notice the brackets which indicate an array is being passed in. Even though this parameter does not have the & symbol between the data type `int` and the name `grades[]`, it is still passed by reference. The third parameter will hold the sum to be computed by the function and is passed by reference as indicated by the & symbol.

Even though all arrays are pass by reference, there are times when we do not want the function to alter the value of the array. Inserting the word **const** before the data type on the formal parameter list prevents the function from altering the array even though it is passed by reference. In the above example the procedure is not really altering the array but using it to get a sum of all its values. The prototype and procedure heading could thus be as follows:

```
void Sumarray(int, const int[], int& ); //prototype without named parameters

void Sumarray(int size, const int grades[], int&  sum) //procedure heading
```

The `Sumarray` procedure could be written as a value returning function as follows:

```
int Sumarray (int size, const int grade[]) //value returning function heading
```

The use of brackets in function prototypes and headings can be avoided by declaring a programmer defined datatype. This is done in the global section with a **typedef** statement.

Example: `typedef int GradeType[50];`

This defines a data type, called `GradeType`, that is an array containing 50 integer memory locations. Since `GradeType` is a data type, it can be used in declaring variables.

```
GradeType grades; // This declares grades as an integer array with 50 elements
```

The following shows the revised code (in bold) of our `Sumarray` example using `typedef`.

```
const int TOTALGRADES = 50; //TOTALGRADES is the maximum size of the array

typedef int GradeType[TOTALGRADES];

//procedure prototype
void Sumarray(int size, const GradeType grades, int&  sum);

int main()
{
  GradeType grades;
   int numberOfRecs, sum;
      :
   Sumarray(numberOfRecs, grades, sum);
      :
   return 0;
}

void Sumarray(int size, const GradeType grades, int&  sum)
{
     sum = 0;
     for (int pos = 0; pos < size; pos++)

        sum = sum + grades[pos];
}
```

Notice that `grades` is now declared as a `GradeType` data type rather than `int grades[]`. `GradeType` has been declared as a "new" data type which is an array of 50 locations, each of which can hold an integer. This method used to eliminate brackets in procedure headings is very useful for multi-dimensional arrays.

Two-Dimensional Arrays

Data is often contained in a table of rows and columns that can be implemented with a two-dimensional array. Suppose we wanted to read data representing profits (in thousands) for a particular year and quarter.

Quarter 1	Quarter 2	Quarter 3	Quarter 4
72	80	10	100
82	90	43	42
10	87	48	53

This can be done using a **two-dimensional array**.

Example:

```
const NO_OF_ROWS = 3;
const NO_OF_COLS = 4;

typedef float ProfitType[NO_OF_ROWS][NO_OF_COLS]; //declares a new data type
                                                  //which is a 2 dimensional
                                                  //array of floats
void main()
{
    ProfitType profit;          // declares profit as a 2 dimensional array

    for (int row_pos = 0; row_pos < NO_OF_ROWS; row_pos++)
      for (int col_pos = 0; col_pos < NO_OF_COLS;  col_pos++)
      {
          cout << "Please input a profit" << endl;
          cin  >> profit[row_pos][col_pos];
      }

}
```

A two dimensional array normally uses two loops (one nested inside the other) to read, process or output data.

How many times will the code above ask for a profit? It processes the inner loop `NO_OF_ROWS * NO_OF_COLS` times, which is 12 times in this case.

Multi-Dimensional Arrays

C++ arrays can have any number of dimensions (although more than three is rarely used). To input, process or output every item in an *n*-dimensional array, you need *n* nested loops.

Arrays of Strings

Any variable declared as **`char`** holds only one character. To hold more than one character in a single variable, that variable needs to be an array of characters. A string (a group of characters that usually form meaningful names or words) is really just an array of characters. A complete lesson on characters and strings is given in Lesson 10.

PRE-LAB WRITING ASSIGNMENT

Fill-in-the-Blank Questions

1. The first subscript of every array in C++ is _______________ and the last is _______________ less than the total number of locations in the array.

2. The amount of memory allocated to an array is based on the ______________ ______________ and the ______________ of locations or size of the array.
3. Array initialization and processing is usually done inside a ______________.
4. The ______________ statement can be used to declare an array type and is often used for multidimensional array declarations so that when passing arrays as parameters, brackets do not have to be used.
5. Multi-dimensional arrays are usually processed within ______________ loops.
6. Arrays used as arguments are always passed by ______________.
7. In passing an array as a parameter to a function that processes it, it is often necessary to pass a parameter that holds the ______________ of ______________ used in the array.
8. A string is an array of ______________.
9. Upon exiting a loop that reads values into an array, the variable used as a(n) ______________ to the array will contain the size of that array.
10. An *n*-dimensional array will be processed within ______________ nested loops when accessing all members of the array.

LESSON 7A

LAB 7.1 Working with One-Dimensional Arrays

Retrieve program `testscore.cpp` from the Lab 7 folder. The code is as follows:

```
// This program will read in a group of test scores( positive integers from 1 to 100)
// from the keyboard.
// The program will calculate the average of all the scores as well as list the highest
// and lowest score
// There will be no more than 100 test scores read in

// PLACE YOUR NAME HERE

#include <iostream.h>

typedef int GradeType[100];  //defining a new data type:  an integer array of 100 elements

float Average (const GradeType, int);   //Finds average of all grades
int   Highest (const GradeType, int);   //Finds highest of all grades
int   Lowest  (const GradeType, int);   //Finds lowest of all grades

void main()

{
   GradeType  grades;                   //The array holding the grades.
   int  sizeOfGrades;                   //The number of grades read.
   int pos;                             //Index to the array.
```

```
    float avgOfGrades;                   //contains the average of the grades.
    int highestGrade;                    //contains the highest grade.
    int lowestGrade;                     //contains the lowest grade.

    //read in the values into the array

    pos = 0;
    cout << "Please input a grade from 1 to 100,  use -99 to stop" << endl;
    cin  >> grades[pos];

    while (grades[pos] != -99)
    {

        // Fill in the code to read the grades

    }

    sizeOfGrades = _________;  //fill in the blank with appropriate identifier

    //call to the function to find average

    avgOfGrades = Average(grades, sizeOfGrades);

    cout << endl << "The average of all the grades is " << avgOfGrades << endl;

    //  Fill in the call to function that calculates highest grade

    cout << endl << "The highest grade is " << highestGrade << endl;

    // Fill in the call to the function that calculates lowest grade
    // Fill in code to write the lowest to the screen

}

float Average (const GradeType  array, int size)

{
    //This function takes an array of integers (along with its size) and
    //returns the average of those numbers

    float sum = 0;    // holds the sum of all the numbers

    for (int pos = 0; pos < size; pos++)

        sum = sum + array[pos];

    return (sum / size);  //returns the average

}

int   Highest (const GradeType array, int size)
```

continues

```
   // This function takes an array of integers and returns the highest value

{

   // Fill in the code for this function

}

int   Lowest  (const GradeType array, int size)

   // This function takes an array of integers and returns the lowest value

{
   // Fill in the code for this function

}
```

Exercise 1: Complete this program by filling in the code (places in bold)

Exercise 2: Run the program with the following data: 90 45 73 21 62 –99. Record the output here:

____

Exercise 3: (Optional) If you have worked with files, change this program so that it reads the information from a file. `gradfile.txt` from the Lab 7 folder has been created with the sample data above. Retrieve it and place it in the same folder as your C++ source code and run the program.

LAB 7.2 Strings as Arrays of Characters

Retrieve program `student.cpp` from the Lab 7 folder.

```
// This program will input an undetermined number of student names and a number
// of grades for each student.
// The number of grades is given by the user.
// The grades are stored in an array.
// Two functions are called for each student.
// One function will give the numeric average of their grades.
// The other function will give a letter grade to that average.
// Grades are assigned on a 10 point spread.
// 90-100 A   80- 89 B  70-79 C   60-69 D   Below 60 F

// PLACE YOUR NAME HERE

#include <iostream.h>

const  int MAXGRADE = 25;  //maximum number of grades per student
const  char MAXCHAR = 30;  //maximum characters used in a name


typedef char StringType30[MAXCHAR + 1];  //character array used for names having 30
                                         //characters or less
typedef  float GradeType[MAXGRADE];      //one dimensional integer array data type
```

```
float GradeAverage(GradeType, int);   //finds grade average by taking array of grades
                                      //and number of grades as parameters

char  LetterGrade(float);             //finds letter grade from average given to it
                                      //as a parameter

void main()

{
    StringType30 firstname, lastname; //two arrays of characters declared
    int numOfGrades;
    GradeType  grades;                //grades is declared as a one dimensional array
    float average;                    //holds the average of a student's grade
    char moreinput;                   //determines if there is more input

     //Input the number of grades for each student

    cout << "Please input the number of grades each student will receive." << endl
         << "This number must be a number between 1 and " << MAXGRADE << " inclusive"
         << endl;

    cin >> numOfGrades;

    while (numOfGrades > MAXGRADE || numOfGrades < 1)
    {
       cout << "Please input the number of grades for each student." << endl
            << "This number must be a number between 1 and " << MAXGRADE << " inclusive"
            << endl;

       cin >> numOfGrades;

    }

     //input names and grades for each student

    cout << "Please input a y if you want to input more students"
       << " any other character will stop the input" << endl;
    cin >> moreinput;

    while ( moreinput == 'y' || moreinput == 'Y')

    {
       cout << "Please input the first name of the student" << endl;
       cin >> firstname;
       cout << endl << "Please input the last name of the student" << endl;
       cin >> lastname;

       for (int count = 0; count < numOfGrades; count++)

       {

          cout << endl << "Please input a grade" << endl;
          // Fill in the input statement to place grade in the array

       }
```

continues

```
        cout << firstname << ' ' << lastname << " has an average of ";

           // Fill in code to get and print average of student to screen
           // Fill in call to get and print letter grade of student to screen

        cout << endl << endl << endl;
        cout << "Please input a y if you want to input more students"
             << " any other character will stop the input" << endl;
        cin >> moreinput;

    }

}

float GradeAverage(GradeType array, int numgrades)

//This function returns the average of the numbers stored in the array
//numgrades contains the number of grades to be processed

{

//Fill in the code for this function

}

char  LetterGrade(float numgrade)

//This function returns the letter grade for the numgrade passed to it
//Grade is based on a 10 point spread

{

//Fill in the code for this function

}
```

Exercise 1: Complete the program by filling in the code. (Areas in bold)

Run the program with 3 grades per student using the sample data below.

Mary Brown	100 90 90
George Smith	90 30 50
Dale Barnes	80 78 82
Sally Dolittle	70 65 80
Conrad Bailer	60 58 71

You should get the following results:

Mary Brown has an average of 93.333 which gives the letter grade of A
George Smith has an average of 56.6667 which gives the letter grade of F
Dale Barnes has an average of 80 which gives the letter grade of B
Sally Dolittle has an average of 71.6667 which gives the letter grade of C
Conrad Bailer has an average of 63 which gives the letter grade of D

LESSON 7B

LAB 7.3 Working with Two-Dimensional Arrays

Look at the following table containing prices of certain items:

12.78	23.78	45.67	12.67
7.83	4.89	5.99	56.84
13.67	34.84	16.71	50.89

These numbers can be read into a two-dimensional array.

Retrieve `price.cpp` from the Lab 7 folder. The code is as follows:

```
// This program will read in prices into a two dim array
// It will print those prices in a table form
// It will determine the highest and lowest price in the array

// PLACE YOUR NAME HERE

#include <iostream.h>
#include <iomanip.h>

const  MAXROWS = 10;
const  MAXCOLS = 10;

typedef float PriceType[MAXROWS][MAXCOLS];          //Creates a new two dim array floating
                                                    //data type

void   GetPrice(PriceType, int&, int&);             //Gets the prices into the array
void   PrintPrice(PriceType, int, int);             //Prints data as a table

void main()

{
   int rowsUsed;                                    //Holds the number of rows used
   int colsUsed;                                    //Holds the number of columns used
   PriceType pricetable;                            //priceTabe is a two dimensional array
                                                    //holding the prices


   GetPrice(pricetable, rowsUsed, colsUsed);        //This calls the procedure that gets
                                                    //the prices into the array
   PrintPrice(pricetable, rowsUsed, colsUsed);      //This calls the procedure that prints
                                                    //the prices of the array

}

void   GetPrice(PriceType table, int& numOfRows, int& numOfCols)

//  This procedure asks the user to input the number of rows and columns used
//  It will then ask the user to input x number of prices  (x = rows * columns)
//  This data is placed in the array

{
```

continues

```
    cout << "Please input the number of rows from 1 to "<< MAXROWS << endl;
    cin >> numOfRows;

    cout << "Please input the number of columns from 1 to "<< MAXCOLS << endl;
    cin >> numOfCols;

    for (int row = 0;  row < numOfRows;  row++)
    {
        for (int col = 0; col < numOfCols;  col++)

            // Fill in the code to read and store the next value
    }
}

void   PrintPrice(PriceType table, int numOfRows, int numOfCols)

// This procedure prints the table of prices

{

    cout << setiosflags(ios::fixed) << setiosflags(ios::showpoint) << setprecision(2);

    for (int row = 0;  row < numOfRows; row++)
    {
        for (int col = 0;  col < numOfCols; col++)

            // Fill in the code to print the table
    }
}
```

Exercise 1: Fill in the code to complete both functions `getPrice` and `printPrice` then run the program with the following data:

```
Please input the number of rows from 1 to 10
2

Please input the number of columns from 1 to 10
3

Please input the price of an item with 2 decimal places
1.45

Please input the price of an item with 2 decimal places
2.56

Please input the price of an item with 2 decimal places
12.98

Please input the price of an item with 2 decimal places
37.86

Please input the price of an item with 2 decimal places
102.34

Please input the price of an item with 2 decimal places
67.89

        1.45    2.56      12.98
        37.86   102.34    67.89
```

Exercise 2: Why does `getPrice` have the parameters `numOfRows` and `numOfCols` passed by reference whereas `printPrice` has those parameters passed by value?

Exercise 3: The following code is a function that returns the highest price in the array. After studying it very carefully, place the function in the above program and have the program print out the highest value.

```
float  HighestPrice(PriceType table, int numOfRows, int numOfCols)
// This function returns the highest price in the array
{

      float highestprice;

      highestprice = table[0][0];  // make first element the highest price

      for (int row = 0;  row < numOfRows;  row++)
            for (int col = 0;  col < numOfCols; col++)
                  if ( highestprice < table[row][col] )

                        highestprice = table[row][col];

      return highestprice;
}
```

NOTE: This is a value returning function. Be sure to include its prototype in the global section.

Exercise 4: Create another value returning function that finds the lowest price in the array and have the program print that value.

Exercise 5: After completing all the exercises above, run the program again with the values from Exercise 1 and record your results.

Exercise 6: (Optional) Look at the following table that contains quarterly sales transactions for three years of a small company. Each of the quarterly transactions are integers (number of sales) and the year is also an integer.

YEAR	Quarter1	Quarter2	Quarter3	Quarter4
2000	72	80	60	100
2001	82	90	43	98
2002	64	78	58	84

We could use a two-dimensional array consisting of 3 rows and 5 columns. Even though there are only four quarters we need 5 columns (the first column holds the year).

Retrieve `quartsal.cpp` from the Lab 7 folder. The code is as follows:

```
// This program will read in the quarter sales transactions for a given number of years
// It will print the year and transactions in a table format
// It will calculate year total transactions as well as quarter total transactions

// PLACE YOUR NAME HERE

#include <iostream.h>
#include <iomanip.h>
```

continues

```
const  MAXYEAR = 10;
const  MAXCOL = 5;

typedef int SalesType[MAXYEAR][MAXCOL];   //Creates a new two dim integer data type

void   GetSales(SalesType, int&);         //Gets the number of sales into the array
void   PrintSales(SalesType, int);        //Prints data as a table
void   TableHeading();                    //Prints table heading

void main()

{
   int yearsUsed;       //Holds the number of years used

   SalesType sales;     //Sales is a two dimensional array holding the sales transactions

   GetSales(sales, yearsUsed);     //This calls the procedure that gets the values into
                                   //the array
   TableHeading();                 //This calls the procedure that prints the Table
                                   //Heading
   PrintSales(sales, yearsUsed);   //This calls the procedure that prints the values of
                                   //the array

}

void TableHeading()

{
   cout << setw(30) << "YEARLY QUARTERLY SALES" << endl << endl
        << endl;

   cout << setw(10) << "YEAR" << setw(10) << "Quarter 1"
        << setw(10) << "Quarter 2" << setw(10) << "Quarter 3"
        << setw(10) << "Quarter 4" << endl;
}

void   GetSales(SalesType  table, int&  numOfYears)

//  This procedure asks the user to input the number of years
//  It will then ask the user to input 4 numbers (sales for 4
//  quarter) for each year
//  This data is placed in the array

{

   cout << "Please input the number of years (1-" << MAXYEAR << ')' << endl;
   cin >> numOfYears;

   // Fill in the code to read and store the next value

}

void   PrintSales(SalesType table, int numOfYears)

// This procedure prints the table

{

   // Fill in the code to print the table

}
```

Fill in the code for both `GetSales` and `PrintSales`.

This is similar to the `price.cpp` program in Exercise 1; however, the code will be different. This is a table that contains something other than sales in column one.

Exercise 7: Run the program so that the chart from Exercise 6 is printed.

LAB 7.4 Student Generated Code Assignments

Option 1: Write the complete age population program given in the Pre-lab Reading Assignment.

Statement of the problem:

Given a list of ages (1 to 100) from the keyboard, the program will tally how many people are in each age group.

Sample Run:

```
Please input an age from one to 100, put -99 to stop
5
Please input an age from one to 100, put -99 to stop
10
Please input an age from one to 100, put -99 to stop
100
Please input an age from one to 100, put -99 to stop
20
Please input an age from one to 100, put -99 to stop
5
Please input an age from one to 100, put -99 to stop
8
Please input an age from one to 100, put -99 to stop
20
Please input an age from one to 100, put -99 to stop
5
Please input an age from one to 100, put -99 to stop
9
Please input an age from one to 100, put -99 to stop
17
Please input an age from one to 100, put -99 to stop
-99

The number of people 5 years old is 3
The number of people 8 years old is 1
The number of people 9 years old is 1
The number of people 10 years old is 1
The number of people 17 years old is 1
The number of people 20 years old is 2
The number of people 100 years old is 1
```

Option 2: Write a program that will input temperatures for consecutive days. The program will store these values into an array and call a function that will return the average of the temperatures. It will also call a function that will return the highest temperature and a function that will return the lowest temperature. The user will input the number of temperatures to be read. There will be no more than 50 temperatures. Use `typedef` to define the array type. The average should be displayed to two decimal places.

Sample Run:

```
Please input the number of temperatures to be read
5
Input temperature 1:
68
Input temperature 2:
75
Input temperature 3:
36
Input temperature 4:
91
Input temperature 5:
84

The average temperature is 70.80
The highest temperature is 91.00
The lowest temperature is 36.00
```

Option 3: Write a program that will input letter grades (A, B, C, D, F), the number of which is input by the user (a maximum of 50 grades). The grades will be read into an array. A function will be called five times (once for each letter grade) and will return the total number of grades in that category. The input to the function will include the array, number of elements in the array and the letter category (A,B, C, D or F). The program will print the number of grades that are A, B, etc.

Sample Run:

```
Please input the number of grades to be read in. (1-50)
6
All grades must be upper case A B C D or F
Input a grade
A
Input a grade
C
Input a grade
A
Input a grade
B
Input a grade
B
Input a grade
D

Number of A=2
Number of B=2
Number of C=1
Number of D=1
Number of F=0
```

LESSON

8 Searching and Sorting Arrays

PURPOSE

1. To introduce the concept of a search routine
2. To introduce the linear and binary searches
3. To introduce the concept of a sorting algorithm
4. To introduce the bubble and selection sorts

PROCEDURE

1. Students should read the Pre-lab Reading Assignment before coming to lab.
2. Students should complete the Pre-lab Writing Assignment before coming to lab.
3. In the lab, students should complete labs assigned to them by the instructor.

Contents	Pre-requisites	Approximate completion time	Page number	Check when done
Pre-lab Reading Assignment		20 min.	116	
Pre-lab Writing Assignment	Pre-lab reading	10 min.	124	
LESSON 8				
Lab 8.1 Working with the Linear Search	Understanding of character arrays	10 min.	125	
Lab 8.2 Working with the Binary Search	Understanding of integer arrays	15 min.	126	
Lab 8.3 Working with Sorts	Understanding of arrays	15 min.	128	
Lab 8.4 Student Generated Code Assignments	Understanding of arrays	20 min.	131	

PRE-LAB READING ASSIGNMENT

Search Algorithms

A search algorithm is a procedure for locating a specific datum from a collection of data.

For example, suppose you want to find the phone number for Wilson Electric in the phonebook. You would open the phonebook to the business section under W and then look for all the entries that begin with the word Wilson. There would most likely be numerous such entries, so you would then look for the one(s) that end with Electric. This is an example of a **search algorithm**. Since each section in the phonebook is alphabetized, this is a particularly easy search. Of course, there are numerous types of "collections of data" that one could search. In this section we will focus on searching arrays. Two algorithms, the linear and binary searches, will be studied. We will see that each algorithm has its advantages and disadvantages.

Linear Search

The easiest array search to understand is probably the **linear search**. This algorithm starts at the beginning of the array and then steps through the elements sequentially until either the desired value is found or the end of the array is reached. For example, suppose we want to find the first occurrence of the letter "o" in the word "harpoon." We can visualize the corresponding character array as follows:

0	1	2	3	4	5	6	7
H	a	r	p	o	o	n	\0

In C++ we can initialize the character array with the desired string:

```
char word[8] = "Harpoon";
```

So `word[0]='H'`, `word[3]= 'p'`, and `word[7] = '\0 '`. The '\0' marks the end of the string and is called the null character. It is discussed further in Lesson 10. If we perform a linear search looking for "o," then we would first check `word[0]` which is not equal to `'o'`. So we would then move to `word[1]` which is also not equal to `'o'`. We continue until we get to `word[4]='o'`. At this point the subscript 4 is returned so we know the position in the array that contains the first occurrence of the letter `'o'`. What would happen if we searched for `'z'`? Certainly we would step through the array until we reached the end and not find any occurrence of `'z'`. What should the search function return in this case? It is customary to return –1 since this is not a valid array subscript. Here is the complete program that performs the linear search:

Sample Program 8.1:

```
// This program performs a linear search on a character array

#include <iostream.h>

int SearchList( char[], int, char); // function prototype
const int SIZE = 8;
```

```
void main()
{
     char word[SIZE] = "Harpoon";
     int found;
     char ch;

     cout << "Enter a letter to search for:" << endl;
     cin >> ch;

     found = SearchList(word, SIZE, ch);

     if (found == -1)
          cout << "The letter " << ch << " was not found in the list" << endl;
     else
          cout << "The letter " << ch << " is in the " << found + 1
               << " position of the list" << endl;

}

int SearchList( char List[], int numElems, char value)
{
     for(int count = 0; count <= numElems; count++) // for loop steps through
                                                    // the array
     {
          if (List[count] == value) // each array entry is checked to see if
                                    // it contains the desired value.
            return count;           // if the desired value is found, the
                                    // array subscript count is returned to
                                    // indicate the location in the array.
     }
     return -1;                     // if the value is never found, -1 is
                                    // returned.
}
```

For example, suppose we wish to search the word "Harpoon" for the letter `'o'`. The function SearchList does the linear search and returns the index 4 of the array where `'o'` is found. However, the program outputs 5 for the position since we want to output the character's position within the string rather than its storage location in the `word` array. You have certainly noticed that there is a second occurrence of `'o'` in the word "Harpoon." However, the linear search does not find it since it quits after finding the first occurrence.

One advantage of the linear search is its simplicity. It is easy to step sequentially through an array and check each element for a designated value. Another advantage is that the elements of the array do not need to be in any order to implement the algorithm. For example, to search the integer arrays

First Array	23	45	12	456	99
Second Array	12	29	45	23	456

for the integer 99, the linear search will work. It will return 4 for the first array and –1 for the second. The main disadvantage of the linear search is that it is time-consuming for large arrays. If the desired piece of data is not in the array, then the search has to check every element of the array before it returns –1. Even if the desired piece of data is in the array, there is a very good chance that a significant portion of the array will need to be checked to find it. So we need a more efficient search algorithm for large arrays.

The Binary Search

A more efficient algorithm for searching an array is the **binary search** which eliminates half of the array every time it does a check. The drawback is that the data in the array must be ordered to use a binary search. If we are searching an array of integers, then the values stored in the array must be arranged in order from largest to smallest or smallest to largest.

Examples: Consider the following three integer arrays:

1)	19	15	13	13	11	6	-1	-3
2)	19	15	16	13	13	11	-1	-3
3)	-3	0	1	1	12	14	18	25

The arrays in 1) and 3) could be searched using a binary search. In 1) the values are arranged largest to smallest and in 3) the values are arranged smallest to largest. However, the array in 2) could not be searched using a binary search due to the first three elements of the array: the values of the elements decrease from 19 to 15 but then increase from 15 to 16.

Now that we know which types of arrays are allowed, let us next describe what the binary search actually does. For the sake of argument, let us assume the values of an integer array are arranged from smallest to largest and the integer we are searching for is stored in the variable `wanted`. We first pick an element in the middle of the array—let us call it `middle`. Think about how the number, whether it be even or odd, of elements in the array affects this choice. If `middle = wanted`, then we are done. Otherwise, `wanted` must be either greater than or less than `middle`. If `wanted < middle`, then since the array is in ascending order we know that `wanted` must be before `middle` in the array so we can ignore the second half of the array and search the first half. Likewise, if `wanted > middle`, we can ignore the first half of the array and search just the second half. In both cases we can immediately eliminate half of the array. Once we have done this, we will choose the middle element of the half that is left over and then repeat the same process until either `wanted` is found or it is determined that `wanted` is not in the array.

The following program performs a binary search on an array of integers that is ordered from largest to smallest. Students should think about the logic of this search and how it differs from the argument given above for data ordered smallest to largest.

Sample Program 8.2:

```
#include <iostream.h>

int BinarySearch(int [], int, int);  // function prototype

const int SIZE = 16;

void main()
{
     int found, value;
     int array[] = {34,19,19,18,17,13,12,12,12,11,9,5,3,2,2,0}; //array to
                                                                 //be searched

     cout << "Enter an integer to search for: " << endl;
     cin >> value;

     found = BinarySearch(array, SIZE, value); //function call to perform the
                                               //binary search on array of
                                               //SIZE elements looking for an
                                               //occurrence of value

     if (found == -1)
          cout << "The value " << value << " is not in the list" << endl;
     else
     {
          cout << "The value " << value << " is in position number "
               << found + 1 << " of the list" << endl;
     }

}

int BinarySearch(int array[], int numElems, int value) //function heading
{
     int first = 0;                //First element of list
     int last = numElems - 1;      //last element of the list
     int middle;                   //variable containing the current middle
                                   //value of the list
     while (first <= last)
     {
          middle = first + (last - first) / 2; //finds the middle element
                                               //of the array
     if (array[middle] == value)
          return middle;                       //if value is in the middle,
                                               //we are done

     else if (array[middle] < value)
          last = middle - 1;                   //toss out the second
                                               //remaining half of the array
                                               //and search the first

     else
          first = middle + 1;                  //toss out the first
                                               //remaining half of the array
                                               //and search the second

     }
     return -1;                                //indicates that value is not
                                               //in the array
}
```

If you run this program and search for 2, the output indicates that 2 is in the 14th position of the array. Since 2 is in the 14th and 15th position, we see that the binary search found the first occurrence of 2 in this particular data set. However, in Lab 8.2 you will search for values other than 2 and see that there are other possibilities for which occurrence of a sought value is found.

Sorting Algorithms

We have just seen how to search an array for a specific piece of data, However, what if we do not like the order in which the data is stored in the array? For example, if a collection of numerical values is not in order, we might like them to be so we can use a binary search to find a particular value. Or, if we have a list of names, we may want them put in alphabetical order. To sort data stored in an array, one uses a **sorting algorithm**. In this section we will consider two such algorithms—the bubble sort and the selection sort.

The Bubble Sort

The bubble sort is a simple algorithm used to arrange data in either ascending (lowest to highest) or descending (highest to lowest) order. To see how this sort works, let us arrange the array below in ascending order.

9	2	0	11	5
`Element 0`	`Element 1`	`Element 2`	`Element 3`	`Element 4`

The bubble sort begins by comparing the first two array elements. If `Element 0 > Element 1`, which is true in this case, then these two pieces of data are exchanged. The array is now the following:

2	9	0	11	5
`Element 0`	`Element 1`	`Element 2`	`Element 3`	`Element 4`

Next elements 1 and 2 are compared. Since `Element 1 > Element 2`, another exchange occurs:

2	0	9	11	5
`Element 0`	`Element 1`	`Element 2`	`Element 3`	`Element 4`

Now elements 2 and 3 are compared. Since 9 < 11, there is no exchange at this step. Next elements 3 and 4 are compared and exchanged:

2	0	9	5	11
`Element 0`	`Element 1`	`Element 2`	`Element 3`	`Element 4`

At this point we are at the end of the array. Note that the largest value is now in the last position of the array. Now we go back to the beginning of the array and repeat the entire process over again. Elements 0 and 1 are compared. Since 2 > 0, an exchange occurs:

0	2	9	5	11
`Element 0`	`Element 1`	`Element 2`	`Element 3`	`Element 4`

Next elements 1 and 2 are compared. Since 2 < 9, no swap occurs. However, when we compare elements 2 and 3 we find that 9 > 5 and so they are exchanged. Since

`Element 4` contains the largest value (from the previous pass), we do not need to make any more comparisons in this pass.

The final result is:

0	2	5	9	11
Element 0	Element 1	Element 2	Element 3	Element 4

The data is now arranged in ascending order and the algorithm terminates. Note that the larger values seem to rise "like bubbles" to the larger positions of the array as the sort progresses.

We just saw in the previous example how the first pass through the array positioned the largest value at the end of the array. This is always the case. Likewise, the second pass will always position the second to largest value in the second position from the end of the array. The pattern continues for the third pass, fourth pass, and so on until the array is fully sorted. Subsequent passes have one less array element to check than their immediate predecessor.

Sample Program 8.3:

```
// This program uses a bubble sort to arrange an array of integers in
// ascending order

#include <iostream.h>

// function prototypes

void BubbleSortArray(int [], int);
void DisplayArray(int[], int);

const int SIZE = 5;

void main()
{
      int values[SIZE] = {9,2,0,11,5};

      cout << "The values before the bubble sort is performed are:" << endl;
      DisplayArray(values,SIZE);

      BubbleSortArray(values,SIZE);

      cout << "The values after the bubble sort is performed are:" << endl;
      DisplayArray(values,SIZE);

}

void DisplayArray(int array[], int elems)     //function heading
{                                             //displays the array
      for (int count = 0; count < elems; count++)
            cout << array[count] << "   " << endl;
}
```

continues

```
void BubbleSortArray(int array[], int elems)
{
     int swap, temp;
     int bottom = elems - 1;        //bottom indicates the end part of the
                                    //array where the largest values have
                                    //settled in order
do
     {
          swap = 0;
          for (int count = 0; count < bottom; count++)
          {
               if (array[count] > array[count+1])
               {                         //the next three lines do a swap
                  temp = array[count];
                  array[count] = array[count+1];
                  array[count+1] = temp;
                  swap = 1;              //indicates that a swap occurred
               }
          }
          bottom--;      //bottom is decremented by 1 since each pass through
                         //the array adds one more value that is set in order

     } while(swap != 0); //loop repeats until a pass through the array when
                         //no swaps occur
}
```

While the bubble sort algorithm is fairly simple, it is inefficient for large arrays since data values only move one at a time.

The Selection Sort

A generally more efficient algorithm for large arrays is the **selection sort**. As before, let us assume that we want to arrange numerical data in ascending order. The idea of the selection sort algorithm is to first locate the smallest value in the array and move that value to the beginning of the array (i.e., position 0). Then the next smallest element is located and put in the second position (i.e., position 1). This process continues until all the data is ordered. An advantage of the selection sort is that for *n* data elements at most *n*-1 moves are required. The disadvantage is that *n(n-1)/2* comparisons are always required. To see how this sort works, let us consider the array we arranged using the bubble sort:

Element 0	Element 1	Element 2	Element 3	Element 4
9	2	0	11	5

First the smallest value is located. It is 0, so the contents of `Element 0` and `Element 2` are swapped:

Element 0	Element 1	Element 2	Element 3	Element 4
0	2	9	11	5

Next we look for the second smallest value. The important point to note here is that we do not need to check `Element 0` again since we know it already contains the smallest data value. So the sort starts looking at `Element 1`. We see that the second smallest value is 2, which is already in `Element 1`. Starting at `Element 2` we see that 5 is the smallest of the remaining values. Thus the contents of `Element 2` and `Element 4` are swapped:

0	2	5	11	9
Element 0	Element 1	Element 2	Element 3	Element 4

Finally, the contents of `Element 3` and `Element 4` are compared. Since 11 > 9, the contents are swapped leaving the array ordered as desired:

0	2	5	9	11
Element 0	Element 1	Element 2	Element 3	Element 4

Sample Program 8.4:

```
#include <iostream.h>

// function prototypes

void SelectionSortArray(int [], int);
void DisplayArray(int[], int);
const int SIZE = 5;

void main()
{
     int values[SIZE] = {9,2,0,11,5};

     cout << "The values before the selection sort is performed are:" <<endl;
     DisplayArray(values,SIZE);

     SelectionSortArray(values,SIZE);
     cout << "The values after the selection sort is performed are:" << endl;
     DisplayArray(values,SIZE);

}

void DisplayArray(int array[], int elems)     // function heading
{                                             // Displays array
     for (int count = 0; count < elems; count++)
          cout << array[count] << "   ";
     cout << endl;
}

void SelectionSortArray(int array[], int elems)
{
     int seek;        //array position currently being put in order
     int minCount;    //location of smallest value found
     int minValue;    //holds the smallest value found

     for (seek = 0; seek < (elems-1); seek++) // outer loop performs the swap
                                              // and then increments seek
```

continues

```
        {
            minCount = seek;
            minValue = array[seek];
            for(int index = seek + 1; index < elems; index++)
              {
                  //inner loop searches through array
                  //starting at array[seek] searching
                  //for the smallest value. When the
                  //value is found, the subscript is
                  //stored in minCount. The value is
                  //stored in minValue.

                if(array[index] < minValue)
                {
                     minValue = array[index];
                     minCount = index;
                 }
              }

                //the following two statements exchange the value of the
                //element currently needing the smallest value found in the
                //pass(indicated by seek) with the smallest value found
                //(located in minValue)

              array[minCount] = array[seek];
              array[seek] = minValue;

        }
}
```

PRE-LAB WRITING ASSIGNMENT

Fill-in-the-Blank Questions

1. The advantage of a linear search is that it is ________________.
2. The disadvantage of a linear search is that it is ________________.
3. The advantage of a binary search over a linear search is that a binary search is ________________.
4. An advantage of a linear search over a binary search is that the data must be ________________ for a binary search.
5. After 3 passes of a binary search, approximately what fraction of the original array still needs to be searched (assuming the desired data has not been found)? ________________
6. While the ________________ sort algorithm is conceptually simple, it can be inefficient for large arrays because data values only move one at a time.
7. An advantage of the ________________ sort is that, for an array of size n, at most $n - 1$ moves are required.
8. Use the bubble sort on the array below and construct the first 3 steps that actually make changes. (Assume the sort is from smallest to largest).

19	-4	91	0	-17
Element 0	Element 1	Element 2	Element 3	Element 4

Element 0	Element 1	Element 2	Element 3	Element 4

Element 0	Element 1	Element 2	Element 3	Element 4

Element 0	Element 1	Element 2	Element 3	Element 4

9. Use the selection sort on the array below and construct the first 3 steps that actually make changes. (Assume the sort if from smallest to largest).

19	-4	91	0	-17
Element 0	Element 1	Element 2	Element 3	Element 4

Element 0	Element 1	Element 2	Element 3	Element 4

Element 0	Element 1	Element 2	Element 3	Element 4

Element 0	Element 1	Element 2	Element 3	Element 4

LESSON 8

LAB 8.1 Working with the Linear Search

Bring in program `linear_search.cpp` from the Lab 8 folder. This is Sample Program 8.1 from the Pre-lab Reading Assignment. The code is the following:

```
// This program performs a linear search on a character array

// PLACE YOUR NAME HERE

#include <iostream.h>

int SearchList( char[], int, char); // function prototype
const int SIZE = 8;

void main()
{
    char word[SIZE] = "Harpoon";
    int found;
    char ch;

    cout << "Enter a letter to search for:" << endl;
    cin >> ch;
```

continues

```
    found = SearchList(word, SIZE, ch);
    if (found == -1)
        cout << "The letter " << ch << " was not found in the list" << endl;
    else
        cout << "The letter " << ch <<" is in the " << found + 1 << " position of the list"
              << endl;

}

int SearchList( char List[], int numElems, char value)
{
    for (int count = 0;count <= numElems; count++)  //for loop steps through the array
    {
        if (List[count] == value)                  //each array entry is checked to see
                                                   //if it contains
                                                   //the desired value.
             return count;                         //if the desired value is found, the
                                                   //array subscript
                                                   //count is returned to indicate the
                                                   //location in the array

    }
    return -1;                                     //if the value is not found, -1 is
                                                   //returned
}
```

Exercise 1: Re-write this program so that it searches an array of integers rather than characters. Search the integer array `nums[8] =`

3	6	-19	5	5	0	-2	99

for several different integers. Make sure you try integers that are in the array and others that are not. What happens if you search for 5?

LAB 8.2 Working with the Binary Search

Bring in program `binary_search.cpp` from the Lab 8 folder. This is Sample Program 8.2 from the Pre-lab Reading Assignment. The code is the following:

```
// This program demonstrates a Binary Search

// PLACE YOUR NAME HERE

#include <iostream.h>

int BinarySearch(int [], int, int);  // function prototype

const int SIZE = 16;

void main()
{
    int found, value;
    int array[] = {34,19,19,18,17,13,12,12,12,11,9,5,3,2,2,0};  // array to be searched
```

```
    cout << "Enter an integer to search for:" << endl;
    cin >> value;

    found = BinarySearch(array, SIZE, value); //function call to perform the binary search
                                              //on array looking for an occurrence of
                                              //value

    if (found == -1)
        cout << "The value " << value << " is not in the list" << endl;
    else
    {
        cout << "The value " << value << " is in position number "
             << found + 1 << " of the list" << endl;
    }

}

int BinarySearch(int array[],int numElems,int value) //function heading
{
    int first = 0;                      //First element of list
    int last = numElems - 1;            //last element of the list
    int middle;                         //variable containing the current middle value of
                                        //the list

    while (first <= last)
    {
        middle = first + (last - first) / 2;   //finds the middle element of the array
    if (array[middle] == value)
        return middle;                         //if value is in the middle, we are done

    else if (array[middle]<value)
        last = middle - 1;                     //toss out the second remaining half of
                                               //the array and search the first

    else
        first = middle + 1;                    //toss out the first remaining half of
                                               //the array and search the second

    }

    return -1;                                 //indicates that value is not in the array
}
```

Exercise 1: The variable `middle` is declared as an integer. The program contains the assignment statement `middle=first+(last-first)/2`. Is the right side of this statement necessarily an integer in computer memory? Explain how the `middle` value is determined by the computer. How does this line of code affect the logic of the program? Remember that `first`, `last`, and `middle` refer to the array positions, not the values stored in those array positions.

Exercise 2: Search the array in the program above for 19 and then 12. Record what the output is in each case.

Note that both 19 and 12 are repeated in the array. Which occurrence of 19 did the search find?

____

Which occurrence of 12 did the search find?

____

Explain the difference.

LAB 8.3 Working with Sorts

Bring in either the program `bubble_sort.cpp` or `selection_sort.cpp` from the Lab 8 folder. These are Sample Programs 8.3 and 8.4, respectively, from the Pre-lab Reading Assignment. The code for both are given below.

bubble_sort.cpp

```
// This program uses a bubble sort to arrange an array of integers in
// ascending order

// PLACE YOUR NAME HERE

#include <iostream.h>

// function prototypes

void BubbleSortArray(int [], int);
void DisplayArray(int[], int);

const int SIZE = 5;

void main()
{
   int values[SIZE] = {9,2,0,11,5};

   cout << "The values before the bubble sort is performed are:" << endl;
   DisplayArray(values,SIZE);

   BubbleSortArray(values,SIZE);

   cout << "The values after the bubble sort is performed are:" << endl;
   DisplayArray(values,SIZE);

}

void DisplayArray(int array[], int elems)     //function heading
{                                             //displays the array
   for (int count = 0; count < elems; count++)
       cout << array[count] << "  " << endl;
}

void BubbleSortArray(int array[], int elems)
{
   int swap, temp;
```

```
    int bottom = elems - 1;       //bottom indicates the end part of the
                                  //array where the largest values have
                                  //settled in order
do
    {
        swap = 0;
        for (int count = 0; count < bottom; count++)
        {
            if (array[count] > array[count+1])
            {                                   //the next three lines do a swap
                temp = array[count];
                array[count] = array[count+1];
                array[count+1] = temp;
                swap = 1;               //indicates that a swap occurred
            }
        }
            bottom--;       //bottom is decremented by 1 since each pass through
                            //the array adds one more value that is set in order

    } while(swap != 0);     //loop repeats until a pass through the array with
                            //no swaps occurs
}
```

selection_sort.cpp

```
// This program uses a selection sort to arrange an array of integers in
// ascending order

// PLACE NAME HERE

#include <iostream.h>

// function prototypes

void SelectionSortArray(int [], int);
void DisplayArray(int[], int);
const int SIZE = 5;

void main()
{
    int values[SIZE] = {9,2,0,11,5};

    cout << "The values before the selection sort is performed are:" << endl;
    DisplayArray(values,SIZE);

    SelectionSortArray(values,SIZE);
    cout << "The values after the selection sort is performed are:" << endl;
    DisplayArray(values,SIZE);

}

void DisplayArray(int array[], int elems)     // function heading

{                                             // Displays array
```

continues

```
    for (int count = 0; count < elems; count++)
        cout << array[count] << "  ";
      cout << endl;
}

void SelectionSortArray(int array[], int elems)

{
    int seek;       //array position currently being put in order
    int minCount;   //location of smallest value found
    int minValue;   //holds the smallest value found

    for (seek = 0; seek < (elems-1);seek++)  // outer loop performs the swap
                                             // and then increments seek
    {
        minCount = seek;
        minValue = array[seek];
        for(int index = seek + 1; index < elems; index++)
            {
              //inner loop searches through array
              //starting at array[seek] searching
              //for the smallest value. When the
              //value is found, the subscript is
              //stored in minCount. The value is
              //stored in minValue.

              if(array[index] < minValue)
              {
                 minValue = array[index];
                 minCount = index;
              }
            }

            //the following two statements exchange the value of the
            //element currently needing the smallest value found in the
            //pass(indicated by seek) with the smallest value found
            //(located in minValue)

        array[minCount] = array[seek];
        array[seek] = minValue;

    }
}
```

Exercise 1: Re-write the sort program you chose so that it orders integers from largest to smallest rather than smallest to largest.

Exercise 2: Modify your program from Exercise 1 so that it prints the array at each step of the algorithm. Try sorting the array

23	0	45	-3	-78	1	-1	9

by hand using whichever algorithm you chose. Then have your program do the sort. Does the output match what you did by hand?

LAB 8.4 Student Generated Code Assignments

Write a program that prompts the user to enter the size of an integer array and then prompts the user to enter the array one element at a time. The program should then perform one of the following algorithms and produce output similar to that of the appropriate program from this lesson:

a. linear search
b. binary search
c. bubble sort
d. selection sort

LESSON SET

9 Pointers

PURPOSE

1. To introduce pointer variables and their relationship with arrays
2. To introduce the dereferencing operator
3. To introduce the concept of dynamic memory allocation

PROCEDURE

1. Students should read the Pre-lab Reading Assignment before coming to lab.
2. Students should complete the Pre-lab Writing Assignment before coming to lab.
3. In the lab, students should complete labs assigned to them by the instructor.

Contents	Pre-requisites	Approximate completion time	Page number	Check when done
Pre-lab Reading Assignment		20 min.	134	
Pre-lab Writing Assignment	Pre-lab reading	10 min.	140	
LESSON 9A				
Lab 9.1 Introduction to Pointer Variables	Basic understanding of pointer variables	15 min.	140	
Lab 9.2 Dynamic Memory	Basic understanding of dynamic memory, new and delete operators	35 min.	141	
LESSON 9B				
Lab 9.3 Dynamic Arrays	Basic understanding of the relationship of pointer variables and arrays	25 min.	143	
Lab 9.4 Student Generated Code Assignments	Basic understanding of pointers, the (*) and (&) symbols, sort and search routines	30 min.	144	

PRE-LAB READING ASSIGNMENT

Pointer Variables

A distinction must always be made between a memory location's address and the data stored at that location. A street address like 119 Main St. is a location that is different than a description of what is at that location: the little red house of the Smith family. So far we have been concerned only with data (the contents of a variable) rather than the address (where in main memory the variable is located). In this lesson we introduce variables that contain an address of a memory location rather than data located at that address. A **pointer** type is a data type that contains the address of a variable. The address of a variable is given by placing the (&) in front of the variable name.

Example: `cout << ∑`

The (&) in front of a variable indicates that the address itself, not the data stored in that location, is the value used. In this case it is the address of the variable `sum` that will be printed to the screen. The address of the variable is shown as a hexadecimal value representing the physical location of the variable. Before this lesson, where have you used this (&) symbol in C++ programming? It was used in the heading of a function (and the prototype) just after the data type describing a pass by reference parameter.

We may declare a variable to be a **pointer**, in which case the values assigned to it must be addresses in computer memory.

Example: `int *ptr;`

The (*) symbol in front of the variable `ptr` indicates that it will hold the address of a memory location. The `int` indicates that the memory location that `ptr` points to will hold integer values. `ptr` is NOT an integer type but rather a pointer that will hold the address of a location in which an integer value can be found. This distinction is most important!

By now there is plenty of confusion between the symbols (*) and (&) and we should pause to discuss their use.

The (&) symbol is basically used on two occasions.

1. The most frequent use we have seen is in the postfix of a type (or prefix of the variable name) of a pass by reference parameter in a function heading/prototype. This is called a reference variable. The memory address of the argument (parameter) is sent to the function instead of the value at that address. When the parameter is used in the function, the compiler automatically **dereferences** the variable. Dereference means that the location of that reference variable (parameter in this case) is accessed to receive or retrieve a value.

   We have looked at the swap function on several occasions. We revisit this routine to show that the (&) symbol is used in the parameters that need to be swapped. The reason is that these values need to be changed by the function and thus we give the address (location in memory) of those values so that the function can change them (swap them).

Example:

```
void swap(int &first, int &second)    //The use of the (&) indicates
                                      //that the parameter is passed by
                                      //reference.
    {
       int temp;

       temp = first;            //since first is a reference variable, the
                                //compiler will automatically access the value
                                //stored there and place it in temp
       first = second;
       second = temp;
    }
```

2. The (&) symbol is also used whenever we are interested in the **address** of a variable rather than its **contents**. This use comes in handy when we are assigning values to pointer variables (discussed later in this section).
   The (*) symbol is also basically used on two occasions. It is used to declare pointer variables. It is also used whenever we are interested in the **contents pointed to by the pointer variable** rather than its address. In this way (*) is used as the **dereferencing operator**.
   In many ways (*) and (&) are the opposite of each other. The (*) symbol is used just before a pointer variable so that we may obtain data rather than the address of a variable. The (&) symbol is used on a non-pointer variable (every variable that we have studied before this lesson) so that an address, rather than the contents of the location, will be used.

Sample Program 9.1:

```
#include <iostream.h>

void main()

{
   int one = 10;                    //one is a "normal" (non pointer) variable
   int* ptr1;                       //ptr1 is a pointer variable that points
                                    //to an int location

   ptr1 = &one;                     //&one indicates that the address, not
                                    //contents, of one is used. That address
                                    //is assigned to the value of ptr1.
                                    //Remember that ptr1, being a pointer
                                    //variable, can only take an address

   cout << "The value of &one is  " << &one << endl;
   cout << "The value of ptr1 is  " << ptr1 << endl;
   cout << "The value of *ptr1 is " << *ptr1 << endl;

}
```

What do you expect will be printed if the address of variable one is the hexadecimal value 006AFOF4?

The following will be printed by the program:

Output	Comments
`The value of &one is 006AF0F4`	**one** is an integer variable, the (&) preceding it returns its address
`The value of ptr1 is 006AF0F4`	**ptr1**, which can only hold an address, was set equal to the address of **one**
`The value of *ptr1 is 10`	The (*) is the dereferencing operator which means ***ptr1** gives us the value of the variable ptr1 is pointing at.

Arrays and Pointers

An array name without any brackets or subscripts is actually a pointer which points to the starting address of the array.

By adding to the name of an array, we can access various elements of that array.

Example:

A	B	C	D	F
starting address	starting address + 1			

The + 1 (above) means 1 times the size of whatever memory allocation is given based on the data type of the array. For example, if a memory location holds 2 bytes then the + 1 means 1 times 2 bytes. The compiler takes care of that so we can think of adding 1 to the starting address as pointing to the next element of the array. Adding 2 would point to the third element, 3 to the fourth element, etc.

```
#include <iostream.h>

void main()
{
   char grades[] = {'A','B','C','D','F'}; //declaring & initializing an array.
                                          //grades is the name of an array and
                                          //so it contains the starting
                                          //address of the array.  The (*) in
                                          //front of grades, dereferences it
                                          //so that the letter A is printed
                                          //instead of the address

   cout << "The first grade is " << *grades << endl;
   cout << "The second grade is " << *(grades + 1) << endl;
   cout << "The third grade is " << *(grades + 2) << endl;
   cout << "The fourth grade is " << *(grades + 3)<< endl;
   cout << "The fifth grade is " << *(grades + 4) << endl;

}
```

What is printed by the program?

```
The first grade is A
The second grade is B
The third grade is C
The fourth grade is D
The fifth grade is F
```

Dynamic Variables

The size of an array has to be given at the time of its declaration. The programmer needs to estimate the maximum number of elements that will be used by the array and this size is static (does not change). As a consequence, if the array is declared to be larger than is needed, memory is wasted. If it is declared to be smaller then is needed, there is not enough memory to hold all the elements.

The use of pointers (and the **new** and **delete** operators described below) allows us to dynamically allocate memory for an array so that memory is not wasted. The following program asks the user to input the number of grades (number of elements that will be used in the array) to be processed.

Pointers allow us to create **dynamic variables** which can be created and destroyed by the program. We have studied scope rules which define where (and when) a variable is active. The scope of dynamic variables is controlled by the program itself by allocating (creating) variables and deallocating (deleting) variables. An operator called **new** is used to allocate and the operator **delete** is used to deallocate dynamic variables.

Example:

```
int *one;               //declares a pointer variable
int *two;
int result;             //declares a variable that will hold the sum of two
                        //values


one = new int;          //dynamically allocates enough memory to hold an int
                        //assigns the address of pointer variable one.

two = new int;
*one = 10;              //assign the value of 10 to the memory location
                        //pointed to by one

*two = 20;

result = *one + *two;
cout << "result = " << result << endl;

delete one;             //deallocates the variable
delete two;
```

Sample Program 9.2:

```
//  This program finds the average of grades
//  It dynamically allocates space for the array holding the grades

#include <iostream.h>

void Sortit(float* grades, int numOfGrades);  //prototype of function to sort

int main()

{
    float *grades;                      //a pointer to an array holding grades
    float total = 0;                    //total of all grades
    float average;                      //average of all grades
    int numOfGrades;                    //The number of grades to be processed
    int count;                          //loop counter

    cout << setiosflags(ios::fixed) << setiosflags(ios::showpoint)
         << setprecision(2);

    cout << "How many grades will be processed " << endl;
    cin >> numOfGrades;

    grades = new float[numOfGrades];  //allocate memory

            //NOTE that new is the operator that is allocating an array of
            //floats that has the number of elements specified by the user.
            //grades is the pointer to the starting address of the array

    if (grades == NULL)    //always test for null

            //A pointer that contains the address 0 is called a null pointer.
            //If grades contained the value of 0 it would indicate that
            //there was not enough memory for the array. An appropriate
            //message, indicating this error should be printed.
            //Null is a reserved word that indicates a non-valid address.

     {
      cout << "Error allocating memory!\n";
      return 1; //Some value other than 0 is returned indicating a problem
     }

    cout<< "Enter the grades below\n";

    for ( count = 0; count < numOfGrades; count++)
      {
          cout << "Grade " << (count+1) << ":";
          cin >> grades[count];
      }
```

```
    for ( count = 0; count < numOfGrades;  count++)
     {
           total = total + grades[count];
     }

    average = total / numOfGrades;

    cout << "Average Grade is " << average << "%" << endl;
    Sortit(grades, numOfGrades);          //call to a sort function
    delete [] grades;                     //deallocates array memory the []
                                          //indicates an array
}

void Sortit(float* grades, int numOfGrades)
{
     //sort code here
}
```

The function `Sortit` is not given in the sample program. Notice how the dynamic array is passed as a parameter. In fact, the pointer variable `grades` is passed and its corresponding formal parameter is declared as a pointer data type.

Dynamic memory is used to save memory space. This is a minor consideration for the type of programs done in this course but a major concern in professional programming environments where large fluctuating amounts of data are used.

Pointers have many practical applications, especially with structures and classes which will be introduced in Lesson Sets 11 and 13.1, respectively.

Review of * and &

When (*) is used in the declaration of a pointer variable (between the data type and the name of the pointer variable), it indicates that the contents of that variable holds an address, NOT the data at that address.

Example 1: `int * ptr1;`

(*) is also used as a dereferencing operator. When placed in front of an already declared pointer variable, the data of the location the pointer points to will be used and NOT the address.

Example 2: `cout << *ptr1;`

Since ptr1 was declared as a pointer variable in Example 1, the output of Example 2 will be the data at that address. The * in this case dereferences the variable `ptr1`.

(&) is used in a procedure or function heading. It is placed between the data type and the parameter name of all parameters that are passed by reference.

(&) is also used before a non-pointer variable to indicate that the address, not the contents, of a location is to be used.

Example 3:

```
int *ptr1;
int one = 10;

ptr1 = &one;                              //This assigns an address of variable
                                          //one to ptr1 because & means "address
                                          //of".
cout <<"The value of &one is  " << &one << endl;   //This prints an address
cout <<"The value of *ptr1 is " << *ptr1 << endl;  //This prints data at that
                                                   //address because (*)is the
                                                   //dereferencing operator.
```

PRE-LAB WRITING ASSIGNMENT

Fill-in-the-Blank Questions

1. The ______________ symbol is the dereferencing operator.
2. The ______________ symbol means "address of."
3. The name of an array, without any brackets, acts as a(n) ______________ to the starting address of the array.
4. An operator that allocates a dynamic variable is ______________.
5. An operator that deallocates a dynamic variable is ______________.
6. A parameter that is passed by ______________ is similar to a pointer variable in that it can contain the address of another variable. They are used as parameters of a procedure (void function) whenever we want a procedure to change the value of the argument.

Given the following information fill the blanks with either "an address" or "3.75".

```
float * pointer;
float pay = 3.75;

pointer = &pay;
```

7. `cout << pointer;` will print ______________.
8. `cout << *pointer;` will print ______________.
9. `cout << &pay;` will print ______________.
10. `cout << pay;` will print ______________.

LESSON 9A

LAB 9.1 Introduction to Pointer Variables

Retrieve program `pointers.cpp` from the Lab 9 folder.
The code is as follows:

```
// This program demonstrates the use of pointer variables
// It finds the area of a rectangle given length and width
// It prints the length and width in ascending order

//PLACE YOUR NAME HERE
```

```
#include <iostream.h>

void main()
{

    int length;              //holds length
    int width;               //holds width
    int area;                //holds area

    int *ptrlength;          //pointer to integer value of length
    int *ptrwidth;           //pointer to integer value of width

    cout << "Please input the length of the rectangle" << endl;
    cin >> length;
    cout << "Please input the width of the rectangle" << endl;
    cin >> width;

    // Fill in code that will give values to ptrlength and ptrwidth.
    // ptrlength will point to the address of the value stored in length
    // ptrwidth will point to the address of the value stored in width

    area = // Fill in code to find the area by using only the pointer variables
    cout << "The area is " << area << endl;

    if (// Fill in the condition of length > width by using only the pointer variables
        cout << "The length is greater than the width" << endl;

    if (// Fill in the condition of width > length by using only the pointer variables)
        cout << "The width is greater than the length" << endl;

    if (// Fill in the condition of length=width by using only the pointer variables
        cout << "The width and length are the same" << endl;

}
```

Exercise 1: Complete this program by filling in the code (places in bold). Note: use only pointer variables when instructed to by the comments in bold. This program is to test your knowledge of pointer variables and the & and * indicators.

Exercise 2: Run the program with the following data: 10 15. Record the output here ____________________

LAB 9.2 Dynamic Memory

Retrieve program `dynamic.cpp` from the Lab 9 folder.

The code is as follows:

```
// This program demonstrates dynamic variables

// PLACE YOUR NAME HERE

#include <iostream.h>

const MAXNAME = 10;

void main()

{

    char * name;
    int * one;
    int * two;
    int * three;
    int result;

//  Fill in code to allocate the integer variable one here
//  Fill in code to allocate the integer variable two here
//  Fill in code to allocate the integer variable three here
//  Fill in code to allocate the character array pointed by name

    cout << "Enter your last name with exactly 10 characters." << endl;
    cout << "If your name < 10, repeat last letter. Blanks do not count." << endl;

    for (int pos = 0;  pos < MAXNAME;   pos++)

       cin >> // Fill in code to read a character into the name array
              // WITHOUT USING a bracketed subscript

    cout << "Hi ";
    for (int pos1 = 0;  pos1 < MAXNAME;  pos1++)

       cout << // Fill in code to a print a character from the name array
               // WITHOUT USING a bracketed subscript

    cout << endl << "Enter three integer numbers separated by blanks" << endl;

    // Fill in code to bring in three numbers.  You are working only with
    // pointer variables

    //echo print
    cout << "The three numbers are " << endl;

    // Fill in code to output those numbers

    result = // Fill in code to generate the sum of the three numbers
    cout << "The sum of the three values is " << result << endl;

    // Fill in code to deallocate one, two, three and name

}
```

Exercise 1: Complete the program by filling in the code. (Areas in bold) This problem requires that you study very carefully the code already written to prepare you to complete the program.

Sample Run:

```
Enter your last name with exactly 10 characters.
If your name < 10 characters, repeat last letter. Blanks do not count.
DeFinooooo
Hi DeFinooooo
Enter three integer numbers separated by blanks
5 6 7
The three numbers are 5 6 7
The sum of the three values is 18
```

Exercise 2: In inputting and outputting the name, you were asked NOT to use a bracketed subscript. Why is a bracketed subscript not necessary?

Would using `name[pos]` work for inputting the name? Why or why not? Would using `name[pos1]` work for outputting the name? Why or why not?

Try them both and see.

LESSON 9B

LAB 9.3 Dynamic Arrays

Retrieve program `darray.cpp` from the Lab 9 folder.

The code is as follows:

```
// This program demonstrates the use of dynamic arrays

//PLACE YOUR NAME HERE

#include <iostream.h>
#include <iomanip.h>

int main()  // NOTE:  Main is now a value returning function

{

    float  *monthsales;       //a pointer to an array holding monthly sales
    float  total = 0;         //total of all sales
    float  average;           //average of monthly sales
    int    numOfSales;        //The number of sales to be processed
    int    count;             //The loop counter

    cout << setprecision(2) << setiosflags(ios::fixed) << setiosflags(ios::showpoint);

    cout << "How many monthly sales will be processed ";
    cin >> numOfSales;

    // Fill in the code to allocate memory for the array pointed to by monthsales.
```

continues

```
    if ( // Fill in the condition to determine if memory has been allocated)

    {
        cout << "Error allocating memory!\n";
        return 1;
    }

    cout << "Enter the sales below\n";

    for ( count = 0; count < numOfSales;  count++)
        {
            cout << "Sales for Month number  " << // Fill in code to show the number
                                                 // of the month << ":";

            // Fill in code to bring sales into an element of the array
        }

    for ( count = 0; count < numOfSales; count++)
        {
            total = total + monthsales[count];
        }

    average = // Fill in code to find the average


    cout << "Average Monthly sale  is $" << average << endl;
    // Fill in the code to deallocate memory assigned to the array.

    return 0;
}
```

Exercise 1: Fill in the code as indicated by the comments in bold.

Sample Run:

```
How many monthly sales will be processed 3
Enter the sales below
Sales for Month number 1: 401.25
Sales for Month number 2: 352.89
Sales for Month number 3: 375.05
Average Monthly sale is $376.40
```

LAB 9.4 Student Generated Code Assignments

In these assignments you are asked to develop functions that have dynamic arrays as parameters. Remember that dynamic arrays are accessed by a pointer variable and thus the parameters that serve as dynamic arrays are, in fact, pointer variables.

Example:

```
void Sort(float* score, int num_scores); //a prototype whose function has a
                                         //dynamic array as its first
                                         //parameter. It is a pointer variable
.
.
int main();
{
   float *score;                         //a pointer variable
     .
     .
   score  =  new float[num_scores];      //allocation of the array

   Sort(score,scoresize);                //call to the function
```

Option 1: Write a program that will read scores into an array. The size of the array should be input by the user (dynamic array). The program will find and print out the average of the scores. It will also call a function that will sort (using a bubble sort) the scores in ascending order. The values are then printed in this sorted order.

Sample Run:

```
Please input the number of scores
5
Please enter a score
100
Please enter a score
90
Please enter a score
95
Please enter a score
100
Please enter a score
90
The average of the scores is 95

Here are the scores in ascending order
90
90
95
100
100
```

Option 2: This program will read in id numbers and place them in an array. The array is dynamically allocated large enough to hold the number of id numbers given by the user. The program will then input an id and call a function to search for that id in the array. It will print whether the id is in the array or not.

Sample Run:

```
Please input the number of id numbers to be read
4
Please enter an id number
96
Please enter an id number
97
Please enter an id number
98
Please enter an id number
99

Please input an id number to be searched
67
67 is not in the array
```

Option 3: Write a program that will read monthly sales into a dynamically allocated array. The program will input the size of the array from the user. It will call a function that will find the yearly sum (the sum of all the sales). It will also call another function that will find the average.

Sample Run:

```
Please input the number of monthly sales to be input
4
Please input the sales for month 1
1290.89
Please input the sales for month 2
905.95
Please input the sales for month 3
1567.98
Please input the sales for month 4
994.83
The total sales for the year is $4759.65
The average monthly sale is $1189.91
```

LESSON

10 Characters and Strings

PURPOSE

1. To demonstrate the unique characteristics of character data
2. To view strings as an array of characters
3. To show how to input and output strings
4. To work with string functions

PROCEDURE

1. Students should read the Pre-lab Reading Assignment before coming to the lab.
2. Students should complete the Pre-lab Writing Assignment before coming to lab.
3. In the lab, students should complete labs assigned to them by the instructor.

Contents	Pre-requisites	Approximate completion time	Page number	Check when done
Pre-lab Reading Assignment		20 min.	148	
Pre-lab Writing Assignment	Pre-lab reading	10 min.	155	
LESSON 10				
Lab 10.1 Character Testing and String Validation	Minimal knowledge of pointers	15 min.	155	
Lab 10.2 Case Conversion	Basic fundamental instructions	10 min.	158	
Lab 10.3 Using `getline()`	Basic knowledge of character arrays	20 min.	159	
Lab 10.4 String Functions—`strcat`	Basic knowledge of character arrays	15 min.	159	
Lab 10.5 Student Generated Code Assignments	Basic knowledge of character arrays	Out of class	160	

PRE-LAB READING ASSIGNMENT

Macros

C++ provides numerous *macros* (small programs) for character testing. These macros will test a single character and return either a non-zero value (true) or zero (false). For example, the macro `isdigit` tests a character to see if it is one of the digits between 0 and 9. So `isdigit(7)` returns a non-zero value whereas `isdigit(y)` and `isdigit($)`both return 0. We will not list all the character macros here. A complete list may be found in the text. The following program demonstrates some of the others. Note that the `ctype.h` header file must be included to use the character macros.

Sample Program 10.1:

```
// This program utilizes several macros for character testing

#include <iostream.h>
#include <ctype.h>

void main()
{
     char input;

     cout << "Please Enter Any Character:" << endl;
     cin >> input;
     cout << "The character entered is " << input << endl << endl;
     cout << "The ASCII code for " << input << " is "
           << int(input) << endl;

     if ( isalpha(input))      // tests to see if the character is a letter
     {
          cout << "The character is a letter" << endl;
          if (islower(input))  // tests to see if letter is lower case
               cout << "The letter is lower case" << endl;
          if (isupper(input))  // tests to see if letter is upper case
               cout << "The letter is upper case" << endl;
     }

     else if (isdigit(input))  // tests to see if character is a digit
          cout << "The character you entered is a digit" << endl;

     else
          cout << "The character entered is neither a letter nor a"
               << "digit" << endl;
}
```

In Lab 10.1 you will see a more practical application of character testing macros.

Character Case Conversion

The C++ library provides the `toupper` and `tolower` functions for converting the case of a character. `toupper` returns the uppercase equivalent for a letter and `tolower` returns the lower case equivalent. For example, `cout << tolower('F');` causes an `f` to be displayed on the screen. If the letter is already lowercase, then `tolower` will return the value unchanged. Likewise, any non-letter argument is returned unchanged by `tolower`. It should be clear to you now what `toupper` does to a given character.

While the `toupper` and `tolower` functions are conceptually quite simple, they may not appear to be very useful. However, the following program shows that they do have beneficial applications.

Sample Program 10.2:

```
// This program shows how the toupper and tolower functions can be
// applied in a C++ program

#include <iostream.h>
#include <ctype.h>
#include <iomanip.h>

void main()
{
     int week, total, dollars;
     float average;
     char choice;

     cout << setiosflags(ios::fixed) << setiosflags(ios::showpoint)
          << setprecision(2);
     do
     {
          total = 0;
          for(week = 1;week <= 4;week++)
          {
               cout << "How much (to the nearest dollar) did you"
                    << " spend on food during week " <<week <<" ?:" <<endl;
               cin >> dollars;
               total = total + dollars;
          }
          average = total / 4.0;

          cout << "Your weekly food bill over the chosen month is $"
               << average << endl << endl;
          do
          {
             cout <<"Would you like to find the average for another month?";
             cout << endl << "Enter Y or N" << endl;
             cin >> choice;
          } while(toupper(choice) != 'Y' && toupper(choice) != 'N');
     } while (toupper(choice) == 'Y');

}
```

This program prompts the user to input weekly food costs, to the nearest dollar (an integer) for a four-week period. The average weekly total for that month is output. Then the user is asked whether they want to repeat the calculation for a different month. The flow of this program is controlled by a `do-while` loop. The condition `toupper(choice) == 'Y'` allows the user to enter `'Y'` or `'y'` for yes. This makes the program more user friendly than if we just allowed `'Y'`. Note the second `do-while` loop near the end of the program. This loop also utilizes `toupper`. Can you determine the purpose of this second loop? How would the execution of the program be affected if we removed this loop (but left in the lines between the curly brackets)?

String Constants

We have already talked about the character data type which includes letters, digits, and other special symbols such as $ and @. Often we need to put characters together to form strings. For example, the price "$1.99" and the phrase "one for the road!" are both strings of characters. The phrase contains blank space characters in addition to letters and an exclamation mark. In C++ a string is treated as a sequence of characters stored in consecutive memory locations. The end of the string in memory is marked by the null character `\0`. Do not confuse the null character with a sequence of two characters (i.e., \ and 0). The null character is actually an escape sequence. Its ASCII code is 0. For example, the phrase above is stored in computer memory as

o	n	e		f	o	r		t	h	e		r	o	a	d	!	\0

A **string constant** is a string enclosed in double quotation marks. For example,

"Learn C++"
"What time is it?"
"Code Word 7dF#c&Q"

are all string constants. When they are stored in the computer's memory, the null character is automatically appended. The string "Please enter a digit" is stored as

P	l	e	a	s	e		e	n	t	e	r		a		d	i	g	i	t	\0

When a string constant is used in C++, it is the memory address that is actually accessed. In the statement

```
cout << "Please enter a digit";
```

the memory address is passed to the `cout` object. `cout` then displays the consecutive characters until the null character is reached.

Storing Strings in Arrays

Often we need to access parts of a string rather than the whole string. For instance, we may want to alter characters in a string or even compare two strings. If this is the case, then a string constant is not what we need. Rather, a character array is the appropriate choice. When using character arrays, enough space to hold the null character must be allocated. For example:

```
char last[10];
```

This code declares a 10-element character array called `last`. However, this array can hold no more than 9 characters since a space is reserved for the null character. Consider the following:

```
char last[10];
cout << "Please enter your last name using no more than 9 letters";
cin >> last;
```

If the user enters `Symon`, then the following will be the contents of the `last` array:

S	y	m	o	n	\0

Recall that the computer actually sees `last` as the beginning address of the array. There is a problem that can arise when using the `cin` object on a character array. `cin` does not "know" that `last` has only 10 elements. If the user enters `Newmanouskous` after the prompt, then `cin` will write past the end of the array. We can get around this problem by using the `getline` function. If we use

```
cin.getline(last,10)
```

then the computer knows that the maximum length of the string, including the null character, is 10. Consequently, `cin` will read until the user hits `ENTER` or until 9 characters have been read, whichever occurs first. Once the string is in the array, it can be processed character by character.

In this next section we will see a program that uses `cin.getline()`.

Library Functions for Strings

The C++ library provides many functions for testing and manipulating strings. For example, to determine the length of a given string one can use the `strlen` function. The syntax is shown in the following code:

```
char line[40] = "A New Day";
int length;
length = strlen(line);
```

Here `strlen(line)` returns the length of the string including white spaces but not the null character at the end. So the value of `length` is 9. Note this is smaller than the size of the actual array holding the string. To use the `strlen` function the `string.h` header file must be included at the top of the program.

To see why we even need a function such as `strlen`, consider the problem of reading in a string and then writing it backwards. If we only allowed strings of a fixed size, say length 29 for example, then the task would be easy. We simply read the string into an array of size 30 or more. Then write the 29th entry followed by the 28th entry and so on, until we reach the 0th entry. However, what if we wish to allow the user to input strings of different lengths? Now it is unclear where the end of the string is. Of course, we could search the array until we find the null character and then figure out what position it is in. But this is precisely what the `strlen` function does for us. Sample Program 10.3 below is a complete program that performs the desired task.

Sample Program 10.3:

```
#include <iostream.h>
#include <string.h>          // necessary to use strlen

void main()
{
     char line[50];
     int length, count = 0;

     cout << "Enter a sentence of no more than 49 characters:\n";
     cin.getline(line,50);

     length = strlen(line);  // strlen returns the length of the string
                             // currently stored in line

     cout << "The sentence you entered printed backwards is:\n";

     for(count = length-1; count >= 0; count--)
     {
          cout << line[count];

     }
     cout << endl;

}
```

Sample Run 1:

```
Enter a sentence of no more than 49 characters:
luaP deiruB I
The sentence you entered printed backwards is:
I Buried Paul
```

Sample Run 2:

```
Enter a sentence of no more than 49 characters:
This sentence is too long to hold a mere 49 characters!
The sentence you entered printed backwards is:
arahc 94 erem a dloh ot gnol oot si ecnetnes sihT
```

Another useful function for strings is `strcat`, which concatenates two strings. `strcat(string1,string2)` attaches the contents of `string2` to the end of `string1`. The programmer must make sure that the array containing `string1` is large enough to hold the concatenation of the two strings plus the null character.

Consider the following code:

```
char string1[25] = "Total Eclipse ";    // note the space after the second
                                        // word - strcat does not insert a
                                        // space. The programmer must do this.
char string2[11] = "of the Sun";

cout << string1 << endl;
cout << string2 << endl;

strcat(string1,string2);

cout << string1 << endl;
```

These statements produce the following output:

```
Total Eclipse
of the Sun
Total Eclipse of the Sun
```

What would have happened if we had declared `string1` to be a character array of size 20?

There are several other string functions such as `strcpy` (copies the second string to the first string), `strcmp` (compares two strings to see if they are the same or, if not, which string is alphabetically greater than the other), and `strstr` (looks for the occurrence of a string inside of another string). For more details on these string functions and the others, see the text.

Pointers and Strings

Pointers can be very useful for writing string processing functions. If one needs to process a certain string, the beginning address can be passed with a pointer variable. The length of the string does not even need to be known since the computer will start processing using the address and continue through the string until the null character is encountered.

Sample Program 10.4 below reads in a string of no more than 50 characters and then counts the number of letters, digits, and whitespace characters in the string. Notice the use of the pointer `StrPtr`, which points to the string being processed. The three functions `countLetters`, `countDigits`, and `countWhiteSpace` all perform basically the same task—the `while` loop is executed until `StrPtr` points to the null character marking the end of the string. In the `countLetters` function, characters are tested to see if they are letters. The `if(isalpha(*StrPtr))` statement determines if the character pointed at by `StrPtr` is a letter. If so, then the counter `occurs` is incremented by one. After the character has been tested, `StrPtr` is incremented by one to test the next character. The other two functions are analogous.

Sample Program 10.4:

```
#include <iostream.h>
#include <ctype.h>

//function prototypes
int countLetters(char*);
int countDigits(char*);
int countWhiteSpace(char*);

void main()
{
     int numLetters, numDigits, numWhiteSpace;
     char inputString[51];

     cout << "Enter a string of no more than 50 characters: " << endl << endl;
     cin.getline(inputString,51);

     numLetters = countLetters(inputString);
     numDigits = countDigits(inputString);
     numWhiteSpace = countWhiteSpace(inputString);
```

continues

```
      cout << "The number of letters in the entered string is "
           << numLetters << endl;
      cout << "The number of digits in the entered string is "
           << numDigits << endl;
      cout << "The number of white spaces in the entered string is "
           << numWhiteSpace << endl;

}

int countLetters(char *StrPtr) //this function counts the number of letters
                               //(both capital and lower case) in the string

{
      int occurs = 0;

      while(*StrPtr != '\0')  //loop is exectued as long as the pointer StrPtr
                              //does not point to the null character which
                              //marks the end of the string
      {
           if (isalpha(*StrPtr))  //isalpha determines if
                                  //the character is a letter
                occurs++;
           StrPtr++;
      }
      return occurs;
}


int countDigits(char *StrPtr)  //this function counts the number of digits
{
      int occurs = 0;

      while(*StrPtr != '\0')
      {
           if (isdigit(*StrPtr))  //isdigit determines if
                                  //the character is a digit
                occurs++;
           StrPtr++;
      }
      return occurs;
}

int countWhiteSpace(char *StrPtr) //this function counts the
                                  //number of whitespace characters.
                                  //These include, space, newline,
                                  //vertical tab, and tab
{
      int occurs = 0;

      while(*StrPtr != '\0')
      {
```

```
        if (isspace(*StrPtr))       //isspace determines if
                                    //the character is a
                                    //whitespace character

            occurs++;
        StrPtr++;
    }
    return occurs;
}
```

PRE-LAB WRITING ASSIGNMENT

Fill-in-the-Blank Questions

1. The code `cout << toupper('b');` causes a _______________ to be displayed on the screen.
2. The data type of `isalpha('g')` is _______________.
3. After the assignment statement `result = isdigit('$')`, `result` has the value _______________.
4. The code `cout << tolower('#');` causes a _______________ to be displayed on the screen.
5. The end of a string is marked in computer memory by the _______________.
6. In `cin.getline(name,25)`, the 25 indicates that the user can input at most _______________ characters into `name`.
7. Consider the following:

   ```
   char message[35] = "Like tears in the rain";
   int length;
   length = strlen(message);
   ```

   Then the value of `length` is _______________.
8. Consider the code

   ```
   char string1[30] = "In the Garden";
   char string2[15] = "of Eden";
   strcat(string1,string2);
   cout << string1;
   ```

   The output for this is _______________.
9. The _______________ header file must be included to access the `islower` and `isspace` character macros.
10. In C++, a string constant must be enclosed in _______________ whereas a character constant must be enclosed in _______________.

LESSON 10

LAB 10.1 Character Testing and String Validation

The American Equities investment company offers a wide range of investment opportunities ranging from mutual funds to bonds. Investors can check the value of their portfolio from the American Equities' web page. Information about personal portfolios is protected via encryption and can only be accessed using a password. The American Equities company requires that a password consist of 8 characters, 5 of which must be letters and the other 3 digits. The letters and digits can be arranged in any order. For example,

rt56AA7q
123actyN
1Lo0Dwa9
myNUM741

are all valid passwords. However, the following are all invalid:

the476NEw	// It contains more than 8 characters (also more than 5 // letters)
be68moon	// It contains less than 3 digits.
$retrn99	// It contains only 2 digits and has an invalid character ('$')

American Equities needs a program for their web page that determines whether or not an entered password is valid. The program `american_equities.cpp` from the Lab 10 folder performs this task. The code is the following:

```
// This program tests a password for the American Equities
// web page to see if the format is correct

// PLACE NAME HERE

#include <iostream.h>
#include <ctype.h>
#include <string.h>

//function prototypes

bool testPassWord(char[]);
int countLetters(char*);
int countDigits(char*);

void main()
{
   char passWord[20];

   cout << "Enter a password consisting of exactly 5 "
        << "letters and 3 digits:" << endl;
   cin.getline(passWord,20);

   if (testPassWord(passWord))
       cout << "Please wait - your password is being verified" << endl;
   else
   {
       cout << "Invalid password. Please enter a password "
            << "with exactly 5 letters and 3 digits" << endl;
       cout << "For example, my37RuN9 is valid" << endl;
   }

}
```

```
bool testPassWord(char custPass[])
{
   int numLetters, numDigits, length;

   length = strlen(custPass);
   numLetters = countLetters(custPass);
   numDigits = countDigits(custPass);
   if (numLetters == 5 && numDigits == 3 && length == 8 )
       return true;
   else
       return false;
}

// the next 2 functions are from Sample Program 10.4

int countLetters(char *StrPtr) //this function counts the
                               //number of letters (both
                               //capital and lower case
                               //in the string

{
   int occurs = 0;

   while(*StrPtr != '\0')

   {
        if (isalpha(*StrPtr))
            occurs++;
       StrPtr++;
   }

   return occurs;
}

int countDigits(char *StrPtr)  //this function counts the
                               //number of digits
{
   int occurs = 0;

   while(*StrPtr != '\0')
   {
       if (isdigit(*StrPtr))  //isdigit determines if
                              //the character is a digit
          occurs++;
       StrPtr++;
   }

   return occurs;
}
```

Exercise 1: Run program american_equities.cpp several times with both valid and invalid passwords. Read through the program and make sure you understand the logic of the code.

Exercise 2: Alter the program so that a valid password consists of 10 characters, 6 of which must be digits and the other 4 letters.

Exercise 3: (Optional) Adjust your program from Exercise 2 so that only lower case letters are allowed for valid passwords.

LAB 10.2 Case Conversion

Bring in case_convert.cpp from the Lab 10 folder. Note that this is Sample Program 10.2. The code is the following:

```
// This program shows how the toupper and tolower functions can be
// applied in a C++ program

// PLACE YOUR NAME HERE

#include <iostream.h>
#include <ctype.h>
#include <iomanip.h>

void main()
{
    int week, total, dollars;
    float average;
    char choice;

    cout << setiosflags(ios::showpoint) << setiosflags(ios::fixed) << setprecision(2);

    do
    {
        total = 0;
        for(week = 1;week <= 4;week++)
        {
            cout << "How much (to the nearest dollar) did you"
                 <<" spend on food during week " << week << " ?:" << endl;
            cin >> dollars;

            total = total + dollars;
        }
        average = total / 4.0;

        cout << "Your weekly food bill over the chosen month is $"
             << average << endl << endl;
        do
        {
           cout << "Would you like to find the average for another month?";
           cout << endl << "Enter Y or N" << endl;
           cin >> choice;

        } while(toupper(choice) != 'Y' && toupper(choice) != 'N');

    } while (toupper(choice) == 'Y');

}
```

Exercise 1: Notice the following `do-while` loop which appears near the end of the program:

```
do
{
   cout << "Would you like to find the average for another month?";
   cout << endl << "Enter Y or N" << endl;
   cin >> choice;
} while(toupper(choice) != 'Y' && toupper(choice) != 'N');
```

How would the execution of the program be different if we removed this loop? Try removing the loop but leave the following lines in the program:

```
cout << "Would you like to find the average for another month?";
cout << endl << "Enter Y or N" << endl;
cin >> choice;
```

Record what happens when you run the new version of the program.

Exercise 2: Alter program `case_convert.cpp` so that it performs the same task but uses `tolower` rather than `toupper`.

LAB 10.3 Using `getline()`

Exercise 1: Write a short program that declares a character array `last` that contains 10 characters. Prompt the user to enter their last name using no more than 9 characters. The program should then read the name into `last` and then output the name back to the screen with an appropriate message.

Exercise 2: Once the program in Exercise 1 is complete, run the program and enter the name **Newmanouskous** at the prompt. What, if anything, happens? (Note that the results could vary depending on your system).

Exercise 3: Re-write the program above using the `getline()` function (and only allowing 9 characters to be input). As before, use the character array `last` consisting of 10 elements. Run your new program and enter **Newmanouskous** at the prompt. What is the output?

LAB 10.4 String Functions—`strcat`

Consider the following code:

```
char string1[25] =" Total Eclipse ";
char string2[11] = "of the Sun";
cout << string1 << endl;
cout << string2 << endl;
strcat(string1,string2);
cout << string1 << endl;
```

Exercise 1: Write a complete program including the above code that outputs the concatenation of `string1` and `string2`. Run the program and record the result.

Exercise 2: Alter the program in Exercise 1 so that `string1` contains 20 characters rather than 25. Run the program. What happens?

LAB 10.5 Student Generated Code Assignments

A **palindrome** is a string of characters that reads the same forwards as backwards. For example, the following are both palindromes:

1457887541 madam

Write a program that prompts the user to input a string of a size 50 characters or less. Your program should then determine whether or not the entered string is a palindrome. A message should be displayed to the user informing them whether or not their string is a palindrome.

LESSON SET

11 Structures and Abstract Data Types

PURPOSE

1. To introduce the concept of an abstract data type
2. To introduce the concept of a structure
3. To develop and manipulate an array of structures
4. To use structures as parameters
5. To use hierarchical (nested) structures

PROCEDURE

1. Students should read the Pre-lab Reading Assignment before coming to lab.
2. Students should complete the Pre-lab Writing Assignment before coming to lab.
3. In the lab, students should complete labs assigned to them by the instructor.

Contents	Pre-requisites	Approximate completion time	Page number	Check when done
Pre-lab Reading Assignment	Chapter 11 of text	20 min.	162	
Pre-lab Writing Assignment	Pre-lab reading	10 min.	171	
LESSON 11 A				
Lab 11.1 Working with Basic Structures	Knowledge of previous chapters	15 min.	171	
Lab 11.2 Initializing Structures	Basic understanding of structures	15 min.	172	
Lab 11.3 Arrays of Structures	Basic understanding of arrays and structures	20 min.	173	
LESSON 11 B				
Lab 11.4 Nested Structures	Basic understanding of functions and nested logic	20 min.	175	
Lab 11.5 Student Generated Code Assignments	Completion of all the previous labs	30 min. + out of class	176	

PRE-LAB READING ASSIGNMENT

So far we have learned of data types such as `float`, `int`, `char`, etc. In some applications the programmer needs to create their own data type. A user defined data type is often called an **abstract data type (ADT)**. The programmer must decide which values are valid for the data type and which operations may be performed on the data type. It may even be necessary for the programmer to design new operations to be applied to the data. We will study this style of programming extensively when we introduce *object-oriented programming* in the lesson sets from Chapter 13.

As an example, suppose you want to create a program to simulate a calendar. The program may contain the following ADTs: `year`, `month`, and `day`. Note that `month` could take on values January, February, . . . , December or even 1,2, . . . ,12 depending on the wishes of the programmer. Likewise, the range of values for `day` could be Monday, Tuesday, . . . , Sunday or even 1,2, . . . ,7. There is much more flexibility in the choice of allowable values for `year`. If the programmer is thinking short term they may wish to restrict `year` to the range 1990–2010. Of course there are many other possibilities.

The example of a C++ abstract data type studied in this lab is the **structure**. Like arrays, structures allow the programmer to group data together. However, unlike an array, structures allow you to group together items of *different* data types. To see how this could be useful in practice, consider what a student must do to register for a college course. Typically, one obtains the current list of available courses and then selects the desired course or courses. The following is an example of a course you may choose:

```
CHEM 310           Physical Chemistry           4 Credits
```

Note that there are four items related to this course: the course discipline (CHEM), the course number (310), the course title (Physical Chemistry), and the number of credit hours (4). We could declare variables as follows:

Variable Declaration	Information Held
`char discipline[4]`	4-letter abbreviation for discipline
`int courseNumber`	integer valued course number
`char courseTitle[20]`	First 20 characters of course title
`short credits`	Number of credit hours

All of these variables are related because they can hold information about the same course. We can package these together by creating a structure. Here is the declaration:

```
struct course
{
     char  discipline[4];
     int   courseNumber;
     char  courseTitle[20];
     short credits;
};   //note the semi-colon here
```

The **tag** is the name of the structure—`course` in this case. The tag is used like a data type name. Inside the braces we have the variable declarations that are the **members** of the structure. So the code above declares a structure named `course` which has four members: `discipline`, `courseNumber`, `courseTitle`, and `credits`.

The programmer needs to realize that the structure declaration above does not define a variable. Rather it lets the compiler know what a `course` structure is composed of. That is, the declaration creates a new data type (an abstract data type) called `course`. We can now declare variables of type `course` as follows:

```
course pChem;
course colonialHist;
```

Both `pChem` and `colonialHist` will contain the four members listed above. We could have also declared these two structure variables on a single line:

```
course pChem, colonialHist;
```

Both `pChem` and `colonialHist` are called **instances** of the `course` structure. In other words, they are both user defined variables that exist in computer memory. Each structure variable contains the four structure members.

Access to Structure Members

Certainly the programmer will need to be able to assign the members values and also keep track of what values the members have. C++ allows you to access structure members using the **dot operator**. Consider the following syntax:

```
colonialHist.credits = 3;
```

In this statement the number 3 is assigned to the `credits` member of `colonialHist`. The dot operator is used to connect the member name to the structure variable it belongs to. To assign our Colonial History course a course discipline, we could use the following code:

```
colonialHist.discipline[0] = H;
colonialHist.discipline[1] = I;
colonialHist.discipline[2] = S;
colonialHist.discipline[3] = T;
```

Now let us put all of these ideas together into a program. Sample Program 11.1 below uses the `course` structure just described. This interactive program allows a student to add requested courses and keeps track of the number of credit hours for which they have enrolled. The execution is controlled by a `do-while` loop.

Sample Program 11.1:

```
#include <iostream.h>
#include <ctype.h>

struct course
{
    char discipline[5];
    int courseNumber;
    char courseTitle[20];
    short credits;
};
```

continues

```
void main()
{
    course nextclass;    // next class is a course structure
    int numCredits = 0;
    char addclass;

    do
    {

        cout << "Please enter course discipline area: ";
        cin >> nextclass.discipline;
        cout << "Please enter the course number: ";
        cin >> nextclass.courseNumber;
        cout << "Please enter the course title: ";
        cin.ignore();    // this is necessary for the next line
        cin.getline(nextclass.courseTitle,20);
        cout << "Please enter the number of credit hours: " << endl;
        cin >> nextclass.credits;

        numCredits = numCredits + nextclass.credits;

          // output the selected course and pertinent information
        cout << "You have been registered for the following: " << endl;
        cout << nextclass.discipline << "      " << nextclass.courseNumber
             << "      " << nextclass.courseTitle
             << "      " << nextclass.credits << " credits" << endl;

        cout << "Would you like to add another class? (Y/N)" << endl;

        cin >> addclass;

    } while(toupper(addclass == 'Y'));

        cout << "The total number of credit hours registered for is: "
             << numCredits << endl;

}
```

Make sure that you understand the logic of this program and, in particular, how structures are used. Notice the line at the end of the `while` loop that reads

```
while(toupper(addclass == 'Y'));
```

What do you think the purpose of `toupper` is?

As a second example, suppose we would like a simple program that computes the area and circumference of two circles input by the user. Although we can easily do this using previously developed techniques, let us see how this can be done using structures. We will also determine which circle's center is further from the origin.

Sample Program 11.2:

```
#include <iostream.h>
#include <math.h>        // necessary for the pow function
#include <iomanip.h>

struct circle          // declares the structure circle. This structure
                       // has 6 members.
{
     float centerx;    // x coordinate of center
     float centery;    // y coordinate of center
     float radius;
     float area;
     float circumference;
     float distance_from_origin;  // distance of center from origin
};

const float PI = 3.14159;

void main()
{
     circle circ1, circ2;  // declares variables circ1 and circ2. The data
                           // type of both is the circle structure

     cout << "Please enter the radius of the first circle: ";
     cin >> circ1.radius;
     cout << "Please enter the x-coordinate of the center: ";
     cin >> circ1.centerx;
     cout << "Please enter the y-coordinate of the center: ";
     cin >> circ1.centery;

     circ1.area = PI * pow(circ1.radius,2.0);
     circ1.circumference = 2 * PI * circ1.radius;
     circ1.distance_from_origin = sqrt(pow(circ1.centerx,2.0) +
            pow(circ1.centery,2.0));

     cout << endl << endl;
     cout << "Please enter the radius of the second circle: ";
     cin >> circ2.radius;
     cout << "Please enter the x-coordinate of the center: ";
     cin >> circ2.centerx;
     cout << "Please enter the y-coordinate of the center: ";
     cin >> circ2.centery;

     circ2.area = PI * pow(circ2.radius,2.0);
     circ2.circumference = 2 * PI * circ2.radius;
     circ2.distance_from_origin = sqrt(pow(circ2.centerx,2.0) +
           pow(circ2.centery,2.0));

     cout << endl << endl;

     // This next part determines which circle's center is closer
     // to the origin
```

continues

```
    if (circ1.distance_from_origin > circ2.distance_from_origin)
    {
          cout << "The first circle is further from the origin"
               << endl << endl;
    }

    else if (circ1.distance_from_origin < circ2.distance_from_origin)
    {
          cout << "The first circle is closer to the origin"
               << endl << endl;
    }
    else
          cout << "The two circles are equidistant from the origin";
          cout << endl << endl;

    cout << setprecision(2);
    cout << setiosflags(ios::fixed) << setiosflags(ios::showpoint);

    cout << "The area of the first circle is :";
    cout << circ1.area << endl;
    cout << "The circumference of the first circle is: ";
    cout << circ1.circumference << endl << endl;

    cout << "The area of the second circle is :";
    cout << circ2.area<<endl;
    cout << "The circumference of the second circle is: ";
    cout << circ2.circumference << endl << endl;

}
```

Arrays of Structures

In the previous sample program we were interested in two instances of the `circle` structure. What if we need a much larger number, say 100, instances of this structure? Rather than declare each one separately, we could use an **array of structures**. An array of structures is declared just like any other array. For example suppose we already have the following structure declaration in our program:

```
struct circle
{
     float centerx;    // x coordinate of center
     float centery;    // y coordinate of center
     float radius;
     float area;
     float circumference;
     float distance_from_origin;   // distance of center from origin
};
```

Then the following statement declares an array, `circn`, which has 100 elements. Each of these elements is a `circle` structure variable:

```
circle circn[100];
```

Like the arrays encountered in previous lessons, you can access an array element using its subscript. So `circn[0]` is the first structure in the array, `circn[1]` is the second and so on. The last structure in the array is `circn[99]`. To access a member of one of these array elements, we still use the dot operator. For instance, `circn[9].circumference` gives the `circumference` member of `circn[9]`. If we wanted to display the center and distance from the origin of the first 30 circles we could use the following:

```
for (int count = 0; count < 30; count++)
{
     cout << circn[count].centerx << endl;
     cout << circn[count].centery << endl;
     cout << circn[count].distance_from_origin;
}
```

When studying arrays you may have seen two-dimensional arrays which allow one to have "a collection of collections" of data. An array of structures allows one to do the same thing. However, we have already noted that structures permit you to group together items of different data type, whereas arrays do not. So using an array of structures can sometimes be used where a two-dimensional array cannot.

Initializing Structures

We have already seen numerous examples of initializing variables and arrays at the time of their declaration in the previous labs. Members of structures can also be initialized when they are declared. Suppose we have the following structure declaration in our program:

```
struct course
{
     char discipline[5];
     int courseNumber;
     char courseTitle[20];
     short credits;
};
```

A structure variable `colonialHist` can be declared and initialized:

```
course colonialHist = {"HIST",302,"Colonial History",3};
```

The values in this list are assigned to `course`'s members in the order they appear. Thus, the string `"HIST"` is assigned to `colonialHist.discipline`, the integer `302` is assigned to `colonialHist.courseNumber`, the string `"Colonial History"` is assigned to `colonialHist.courseTitle`, and the short value `3` is assigned to `colonialHist.credits`. It is not necessary to initialize all the members of a structure variable. For example, we could initialize just the first member:

```
course colonialHist = {"HIST"};
```

This statement leaves the last three members uninitialized. We could also initialize only the first two members:

```
course colonialHist = {"HIST",302};
```

There is one important rule, however, when initializing structure members. If one structure member is left uninitialized, then all structure members that follow it must be uninitialized. In other words, we cannot skip members of a structure during the initialization process.

It is also worth pointing out that you cannot initialize a structure member in the declaration of the structure. The following is an illegal declaration:

```
// illegal structure declaration
struct course
{
     char discipline[5] = "HIST";                  // illegal
     int courseNumber = 302;                       // illegal
     char courseTitle[20] = "Colonial History";    // illegal
     short credits = 3;                            // illegal
};
```

If we recall what a structure declaration does, it is clear why the above code is illegal. A structure declaration simply lets the compiler know what a structure is composed of. That is, the declaration creates a new data type (called `course` in this case). So the structure declaration does not define any variables. Hence there is nothing that can be initialized there.

Hierarchical (Nested) Structures

Often it is useful to nest one structure inside of another structure. Consider the following:

Sample Program 11.3:

```
#include <iostream.h>
#include <iomanip.h>
#include <math.h>

struct center_struct
{
     float x;      // x coordinate of center
     float y;      // y coordinate of center
};

struct circle
{
     float radius;
     float area;
     float circumference;
     center_struct coordinate;
};

const float PI = 3.14159;

void main()
{
```

```
    circle circ1, circ2;  // declares variables circ1 and circ2.
                          // The data type of
                          // both is the circle structure

    cout << "Please enter the radius of the first circle: ";
    cin >> circ1.radius;
    cout << "Please enter the x-coordinate of the center: ";
    cin >> circ1.coordinate.x;
    cout << "Please enter the y-coordinate of the center: ";
    cin >> circ1.coordinate.y;


    circ1.area = PI * pow(circ1.radius,2.0);
    circ1.circumference = 2 * PI * circ1.radius;

    cout << endl << endl;

    cout << "Please enter the radius of the second circle: ";
    cin >> circ2.radius;
    cout << "Please enter the x-coordinate of the center: ";
    cin >> circ2.coordinate.x;
    cout << "Please enter the y-coordinate of the center: ";
    cin >> circ2.coordinate.y;

    circ2.area = PI * pow(circ2.radius,2.0);
    circ2.circumference = 2 * PI * circ2.radius;

    cout << endl << endl;

    cout << setprecision(2);
    cout << setiosflags(ios::fixed) << setiosflags(ios::showpoint);

    cout << "The area of the first circle is :";
    cout << circ1.area << endl;
    cout << "The circumference of the first circle is: ";
    cout << circ1.circumference << endl;
    cout << "Circle 1 is centered at ("<< circ1.coordinate.x << ","
         << circ1.coordinate.y << ")." << endl << endl;

    cout << "The area of the second circle is :";
    cout << circ2.area << endl;
    cout << "The circumference of the second circle is: ";
    cout << circ2.circumference << endl;
    cout << "Circle 2 is centered at (" << circ2.coordinate.x << ","
         << circ2.coordinate.y << ")." << endl << endl;

}
```

Structures and Functions

Just as we can use other variables as function arguments, structure members may be used as function arguments. Consider the following structure declaration:

```
struct circle
{
     float centerx;    // x coordinate of center
     float centery;    // y coordinate of center
     float radius;
     float area;
};
```

Suppose we also have the following function definition in the same program:

```
float computeArea(float r)
{
      return PI * r * r;   // PI must previously be declared as a
                           // constant float
}
```

Let `firstCircle` be a variable of the `circle` structure type. The following function call will pass `firstCircle.radius` into `r`. The return value will be stored in `firstCircle.area`:

```
firstCircle.area = computeArea(firstCircle.radius);
```

It is also possible to pass an entire structure variable into a function rather than an individual member.

```
struct course
{
      char discipline[5];
      int courseNumber;
      char courseTitle[20];
      short credits;
};

course pChem;
```

Suppose the following function definition uses a `course` structure variable as its parameter:

```
void displayInfo(course C)
{
     cout << C.discipline << endl;
     cout << C.courseNumber << endl;
     cout << C.courseTitle << endl;
     cout << C.credits << endl;
}
```

Then the following call passes the `pChem` variable into C:

```
displayInfo(pChem);
```

There are many other topics relating to functions and structures such as returning a structure from a function and pointers to structures. Although we do not have time to develop these concepts in this lab, the text does contain detailed coverage of these topics for the interested programmer.

PRE-LAB WRITING ASSIGNMENT

Fill-in-the-Blank Questions

1. The name of a structure is called the ______________.
2. An advantage of structures over arrays is that structures allow one to use items of ______________ data types, whereas arrays do not.
3. One structure inside of another structure is an example of a ______________.
4. The variables declared inside the structure declaration are called the ______________ of the structure.
5. When initializing structure members, if one structure member is left uninitialized, then all the structure members that follow must be ______________.
6. Another name for a user defined data type is ______________.
7. Once an array of structures has been defined, you can access an array element using its ______________.
8. The ______________ allows the programmer to access structure members.
9. You may not initialize a structure member in the ______________.
10. Like variables, structure members may be used as ______________ arguments.

LESSON 11 A

LAB 11.1 Working with Basic Structures

Bring in program `rect_struct.cpp` from the Lab 11 folder. The code is shown below.

```
// This program uses a structure to hold data about a rectangle

// PLACE YOUR NAME HERE

#include <iostream.h>
#include <iomanip.h>

// Fill in code to declare a structure named rectangle which has members length, width,
// area, and perimeter all of which are floats

void main()
{
    // Fill in code to declare a rectangle variable named box

    cout << "Enter the length of a rectangle: ";

    // Fill in code to read in the length member of box

    cout << "Enter the width of a rectangle: ";

    // Fill in code to read in the width member of box

    cout << endl << endl;
```

continues

```
    // Fill in code to compute the area member of box
    // Fill in code to compute the perimeter member of box

    cout << setiosflags(ios::fixed) << setiosflags(ios::showpoint) << setprecision(2);

    // Fill in code to output the area with an appropriate message
    // Fill in code to output the perimeter with an appropriate message

}
```

Exercise 1: Fill in the code as indicated by the comments in bold.

Exercise 2: Add code to the program above so that the modified program will determine whether or not the rectangle entered by the user is a square.

Sample Run:

```
Enter the length of a rectangle: 7
Enter the width of a rectangle: 7
The area of the rectangle is 49.00
The perimeter of the rectangle is 28.00
The rectangle is a square.
```

LAB 11.2 Initializing Structures

Bring in program `init_struct.cpp` from the Lab 11 folder. The code is shown below.

```
// This program demonstrates partially initialized strucure variables

// PLACE YOUR NAME HERE

#include <iostream.h>
#include <iomanip.h>

struct taxPayer
{
    char name[25];
    long int socialSecNum;
    float taxRate;
    float income;
    float taxes;
};

void main()
{
    // Fill in code to initialize a structure variable named citizen1 so that
    // the first three members are initialized. Assume the name is Tim McGuiness,
    // the social security number is 255871234, and the tax rate is .35

    // Fill in code to initialize a structure variable named citizen2 so that the first
    // three members are initialized. Assume the name is John Kane, the social security
    // number is 278990582, and the tax rate is .29
```

```
    cout << setprecision(2) << setiosflags(ios::fixed) << setiosflags(ios::showpoint);

    // calculate taxes due for citizen1

    // Fill in code to prompt the user to enter this year's income for the tax payer
    // read in this income to the appropriate structure member

    // Fill in code to determine this year's taxes for the tax payer

    cout << "Name: " << citizen1.name << endl;
    cout << "Social Security Number: " << citizen1.socialSecNum << endl;
    cout << "Taxes due for this year: " << citizen1.taxes << endl << endl;

    // calculate taxes due for citizen2

    // Fill in code to prompt the user to enter this year's income for the tax payer
    // read in this income to the appropriate structure member

    // Fill in code to determine this year's taxes for the tax payer

    cout << "Name: " << citizen2.name << endl;
    cout << "Social Security Number: " << citizen2.socialSecNum << endl;
    cout << "Taxes due for this year: " << citizen2.taxes << endl << endl;

}
```

Exercise 1: Fill in the code as indicated by the comments in bold.

Sample Run:

```
Please input the yearly income for Tim McGuiness: 30000
Name: Tim McGuiness
Social Security Number: 255871234
Taxes due for this year: $10500.00

Please input the yearly income for John Kane: 60000
Name: John Kane
Social Security Number: 278990582
Taxes due for this year: $17400.00
```

LAB 11.3 Arrays of Structures

Bring in program `array_struct.cpp` from the Lab 11 folder. The code is shown below.

```
// This program demonstrates how to use an array of strucures

// PLACE YOUR NAME HERE

// Fill in code to put the necessary include statement here for cout and cin
// Fill in code to put any other include statement that the program may need

// Fill in code to declare a structure called taxPayer that has three members:
// taxRate, income, and taxes -- each of type float
```

continues

```
void main()
{
    // Fill in code to declare an array named citizen which holds 5 taxPayer strucures

    cout << setprecision(2) << setiosflags(ios::fixed) << setiosflags(ios::showpoint);


    cout << "Please enter the annual income and tax rate for 5 tax payers: " << endl;
    cout << endl << endl;

    for(int count = 0;count < 5;count++)
    {

        cout << "Enter this year's income for tax payer " << (count + 1);
        cout << ": ";

        // Fill in code to read in the income to the appropriate place

        cout << "Enter the tax rate for tax payer # " << (count + 1);
        cout << ": ";

        // Fill in code to read in the tax rate to the appropriate place

        // Fill in code to compute the taxes for the citizen and store it
        // in the appropriate place

        cout << endl;
    }

    cout << "Taxes due for this year: " << endl << endl;

    // Fill in code for the first line of a loop that will output the tax information
    {
        cout << "Tax Payer # " << (index + 1) << ": " << "$ "
             << citizen[index].taxes << endl;
    }

}
```

Exercise 1: Fill in the code as indicated by the comments in bold.

Exercise 2: In the code above we have the following:

```
cout << "Tax Payer # " << (index+1) << ": " << "$ "
     << citizen[index].taxes << endl;
```

Why do you think we need `(index+1)` in the first line but `index` in the second? Why should these correspond in the program?

Sample Run:

Enter this year's income for tax payer 1: 45000
Enter the tax rate for tax payer # 1: .19
Enter this year's income for tax payer 2: 6000
Enter the tax rate for tax payer # 2: .23
Enter this year's income for tax payer 3: 12000
Enter the tax rate for tax payer # 3: .01

```
Enter this year's income for tax payer 4: 104000
Enter the tax rate for tax payer # 4: .30
Enter this year's income for tax payer 5: 50000
Enter the tax rate for tax payer # 5: .22

Tax Payer # 1: $ 8550.00
Tax Payer # 2: $ 13800.00
Tax Payer # 3: $ 120.00
Tax Payer # 4: $ 31200.00
Tax Payer # 5: $ 11000.00
```

LESSON 11 B

LAB 11.4 Nested Structures

Bring in program `nestedRect_struct.cpp` from the Lab 11 folder. This code is very similar to the rectangle program from Lab 11.1. However, this time you will complete the code using nested structures. The code is shown below.

```
// This program uses a structure to hold data about a rectangle

// PLACE YOUR NAME HERE

#include <iostream.h>
#include <iomanip.h>

// Fill in code to declare a structure named dimensions that contains
// 2 float members, length and width

// Fill in code to declare a structure named rectangle that contains 3 members,
// area, perimeter, and sizes. area and perimeter should be floats, whereas sizes
// should be a dimensions structure variable

void main()
{
    // Fill in code to declare a rectangle structure variable named box.

    cout << "Enter the length of a rectangle: ";

    // Fill in code to read in the length to the appropriate location

    cout << "Enter the width of a rectangle: ";

    // Fill in code to read in the width to the appropriate location

    cout << endl << endl;

    // Fill in code to compute the area and store it in the appropriate location
    // Fill in code to compute the perimeter and store it in the appropriate location

    cout << setiosflags(ios::fixed) << setiosflags(ios::showpoint) << setprecision(2);
    cout << "The area of the rectangle is " << // Fill in code here
    cout << "The perimeter of the recangle is " <<  // Fill in code here

}
```

Exercise 1: Fill in the code as indicated by the comments in bold.

Exercise 2: Modify the program above by adding a third structure named `results` which has two members `area` and `perimeter`. Adjust the `rectangle` structure so that both of its members are structure variables.

Exercise 3: Modify the program above by adding functions that compute the area and perimeter. The structure variables should be passed as arguments to the functions.

Sample Run:

```
Enter the length of a rectangle: 9
Enter the width of a rectangle: 6
The area of the rectangle is 54.00
The perimeter of the rectangle is 30.00
```

LAB 11.5 Student Generated Code Assignments

Option 1: Re-write Sample Program 11.2 so that it works for an array of structures. Write the program so that it compares 6 circles. You will need to come up with a new way of determining which circle's center is closest to the origin.

Option 2: Write a program that uses a structure to store the following information for a particular month at the local airport:

Total number of planes that landed
Total number of planes that departed
Greatest number of planes that landed in a given day that month
Least number of planes that landed in a given day that month

The program should have an array of twelve structures to hold travel information for the entire year. The program should prompt the user to enter data for each month. Once all data is entered, the program should calculate and output the average monthly number of landing planes, the average monthly number of departing planes, the total number of landing and departing planes for the year, and the greatest and least number of planes that landed on any one day (and which month it occurred in).

LESSON SET

12 Files

PURPOSE

1. To introduce the concept of files
2. To use files for I/O
3. To understand and use various types of files (binary and text)

PROCEDURE

1. Students should read the Pre-lab Reading Assignment before coming to lab.
2. Students should complete the Pre-lab Writing Assignment before coming to lab.
3. In the lab, students should complete labs assigned to them by the instructor.

PRE-LAB READING ASSIGNMENT

Introduction to Files

A **file** is a collection of information stored (usually) on a disk. So far most of the programs developed have been interactive: data was input from the keyboard and output was sent to the screen. However, this is inefficient for large amounts of data. Many businesses require programs that get their input from files. Payroll programs, for example, generally read employee records containing such data as hours worked and pay rate, from a file. The derived information (net and gross pay) may be sent to another file to be printed at a later date. Many such non-interactive programs are used in business and scientific applications. This set of labs introduces the concept of files and shows their operations in C++.

There are naming standards for files based on the operating system used. Most conventions allow a name followed by a maximum three letter extension that describes the type of file.

Example:	`auntsusie.wpd`	A word processing file
	`payroll.cpp`	A C++ source program file
	`payroll.exe`	An executable file

A file must first be created and then opened. Then either data is read from a file or information (processed data) is written to a file (or both). Finally, the file is closed when the program is finished.

Setting up a File in C++

We have been using `cin` to get data from the keyboard and `cout` to send information to the screen. Before using either instruction, we first had to use the `#include <iostream.h>` directive. When using files we must use the `#include <fstream.h>` directive.

Files, just like variables, have to be declared in the program. The data type of a file depends on whether it is used as an input file, output file, or both. Output files have a data type called `ofstream`, input files have a data type of `ifstream` and files used as both have the data type `fstream`.

Examples:

```
ofstream outfile;    // declaring outfile as an output file
ifstream infile;     // declaring infile as an input file
fstream datafile;    // declaring datafile to be both an input and output file
```

After declaration, files must still be opened, used (information stored to or data read from the file), and then closed. A file is opened with the `open` function. This ties the logical name of the file that is used in the declaration statement to the physical name of the file used in the secondary storage device (disk). The statement `infile.open("payroll.dat");` opens the file `payroll.dat` and lets the program know that `infile` is the name by which this file will be referenced within the program. If the file is not located in the same directory as the C++ program, the full path (drive, etc.) MUST be indicated: `infile.open("a:\\payroll.dat");`

This tying of the logical name `infile` with the physical name `payroll.dat` means that wherever `infile` is used in the program, data will be read from the physical file `payroll.dat`. There is a problem in opening input files if the physical file (`payroll.dat` in this case) does not exist. A program should check to make sure the `open` function worked. This is done simply by using a conditional statement to check if the file could be found. The following example illustrates how this is done.

Example:

```
ifstream infile;
infile.open("payroll.dat");
if (!infile)                        // or   if (infile == 0)
{
cout << "Error opening file.\n";
cout << "It may not exist where";
cout << " indicated.\n";
return 1;
}
```

Note: In the above example, `return 1` is used as an indicator of an abnormal occurrence. In this case the file in question could not be found.

Output files are opened the same way: `outfile.open("payment.out")`. Whenever the program writes to `outfile`, the information will be placed in the physical file `payment.out`.

Example:

```
ofstream outfile;
outfile.open("info.out");
```

Usage of Files

Just as `cin >>` is used to read from the keyboard and `cout <<` is used to write to the screen, the name of the file, such as `filename >>`, is used to read from the file and `filename <<` is used to write to the file.

Example:

```
infile >> name >> rate >> hours;
outfile << netpay;
```

Closing a File

Files should be closed before the program ends.

```
infile.close();
outfile.close();
datafile.close();
```

Sample Program 12.1:

```
#include <fstream.h>  // library to handle files
#include <iomanip.h>  // library to handle output formatting
int main()
{
     ifstream infile;            // declaration of an input file
     ofstream outfile;           // declaration of an output file
     int hours;                  // the number of hours worked
     float payRate;              // the rate per hour paid
     float net;                  // the net pay
```

continues

```
     infile.open("payroll.dat");  // open infile as an input file
                                  // Whenever infile is used, data from the
                                  // file payroll.dat will be read.
     outfile.open("payment.out"); // open outfile as an output file
                                  // Whenever outfile is used information
                                  // will be set to the file payment.out.

     if(!infile)                  // determines if the input file exists
     {
        cout << "Error opening file.\n";
        cout << "Perhaps the file is not where indicated.\n";
        return 1;
     }
     outfile << setiosflags(ios::fixed) << setprecision(2);
                                  // set format for the output file

     infile >> hours >> payRate;
     net = hours * payRate;
     outfile << "Hours    Pay Rate     Net Pay" << endl;
     outfile << hours << setw(10) << '$' << payRate << setw(9) << '$' << net;

     infile.close();
     outfile.close();

     return 0;
}
```

Assuming that `payroll.dat` contains the values of `40` and `10.00` (in this order), the following will be sent to the file `payment.out`.

```
Hours    Pay Rate    Net Pay
40       $10.00      $400.00
```

File Formatting

Notice the statement `outfile << setiosflags(ios::fixed) << setprecision(2);` in the above example. This shows that the format procedures learned for `cout` can be used for output files as well. The `setw(x)` statement can also be used with output files as long as the `iomanip.h` directive is included.

Reading from a File

Files have an "invisible" end of line marker at the end of each line of the file. Files also have an "invisible" end of file marker at the end of the file. When reading from an input file within a loop, the program must be able to detect that marker as the sentinel data (data that meets the condition to end the loop). You will learn several ways to do this.

Sample Program 12.2:

```
#include <fstream.h>
#include <iostream.h>
#include <iomanip.h>
```

```
int main()
{
    ifstream infile;               // declaration of an input file
    ofstream outfile;              // declaration of an output file

    infile.open("payroll.dat");    // open infile as an input file
                                   // Whenever infile is used, data from the
                                   // file payroll.dat will be read.
    outfile.open("payment.out");   // open outfile as an output file
                                   // Whenever outfile is used information will
                                   // be set to the file payment.out.

    int hours;                     // the number of hours worked
    float payRate;                 // the rate per hour paid
    float net;                     // net pay

    if(!infile)
    {
       cout << "Error opening file.\n";
       cout << "Perhaps the file is not where indicated.\n";

       return 1;
    }

    outfile << setiosflags(ios::fixed) << setprecision(2);
    outfile << "Hours    Pay Rate     Net Pay" << endl;

    infile >> hours;

    while(infile)    // as long as data is in the file the loop will continue
    {
       infile >> payRate;
       net = hours * payRate;

       outfile << hours << setw(10) << "$ " << setw(6) << payRate << setw(5)
               << "$ " << setw(7) << net << endl;
       infile >> hours;
    }
    infile.close();
    outfile.close();

    return 0;
}
```

This program assumes that a data file exists that contains an undetermined number of records with each record consisting of two data values, `hours` and `payrate`. Suppose that the input data file (`payroll.dat`) contains the following:

```
40  10.00
30   6.70
50  20.00
```

The program will produce the following output file (payment.out).

```
Hours      Pay Rate    Net Pay
40          $ 10.00    $ 400.00
30          $  6.70    $ 201.00
50          $ 20.00    $1000.00
```

The input file contains data for one employee on each line. Each time through the `while` loop, information is processed for one employee. The loop executes the same number of times as there are lines (employee records in this case) in the data file. Since there are two items of data for each line (`hours` and `payRate`), these items are read in each time through the loop. Notice that one of the input variables was input before the `while` loop. This is called "priming the read." Input can be thought of as a stream of values taken one at a time. Before the while loop condition can be tested, there has to be something in that "stream." We **prime** the read by reading in at least one variable before the loop. Notice that the statement `infile >> hours;` is given twice in the program: once before the input loop and as the last statement in the loop. The other item, `payrate,` was read in at the very beginning of the loop. In this way each variable will be read each time through the loop. Also notice that the heading of the output file is printed outside the loop before it is executed.

There are other ways of determining that the end of the file has been reached. The `eof()` function reports when the end of a file has been encountered. The above loop could be written as the following:

```
infile >> hours;
while (!infile.eof())
   {
      infile >> payRate;
      net = hours * payRate;
      outfile << hours << setw(10) << "$ " << setw(6) << payRate << setw(5)
              << "$ " << setw(7) << net << endl;
      infile >> hours;
   }
```

In addition to checking to see if a file exists, we can also check to see if it has any data in it. The following code checks first if the file exists and then if it is empty.

```
inData.open("sample2.dat");

if(!inData)
   cout << "file does not exist" << endl;
else if(ch = inData.peek()) == EOF)

   cout << "File is empty" << endl;
else
    //rest of program
```

The `peek` function actually looks ahead in the file for the next data item, in this case to determine if it is the end of file marker. `ch` must be declared as `char` data type.

Since the peek function looks "ahead" for the next data item, it could be used to test for end of file in reading values from a file within a loop without priming the read.

Passing Files as Parameters to Functions

Files can be passed as parameters just like variables. Files, like arrays, are always passed by reference. Unlike arrays however, the & must be included after the data type in the function heading and prototype.

Example:

```
void GetData(ifstream& infile, ofstream& outfile); // prototype of function
                                                   // with files as parameters

void GetData(ifstream& infile, ofstream& outfile)  // heading of function with
                                                   // files as parameters
```

Character Input

Recall that each file has an end of line marker for each line as well as the end of file marker at the end of the file. Whenever whitespace (blanks, newlines, controls, etc.) is part of a file, a problem exists with the traditional >> operator in inputting character data. When reading input characters into a string object, the >> operator skips any leading whitespace. It then reads successive characters into the character array, stopping at the first trailing whitespace character (which is NOT "consumed," but rather which remains as the next character to be read in the file). The >> also takes care of adding the null character to the end of the string. Stopping at the first trailing whitespace can create a problem for names containing spaces. A program reading first names into some string variable (array of characters) would have a problem reading a name like Mary Lou since it has a blank space in it. The blank space between Mary and Lou causes the input to stop when using the >> operator. The get function can be used to input such strings.

```
infile.get(firstname, 20);
```

The get function does NOT skip leading whitespace characters but rather continues to read characters until it either has read, in the example above, 19 characters or it reaches the newline character \n (which it does NOT consume). Recall from Lesson 10 that the last space is reserved for the null character.

Since the get function does not consume the end of line character, there must be something done to consume it so that a new line can be read.

Example: Given the following data file

```
Mary Lou <eol>
Becky    <eol>
Debbie   <eol>
<eof>
```

Note: Both the <eol> and <eof> are NOT visible to the programmer or user.

There are several options for reading and printing this data.

```
char dummy;                  //created to read the end of line character
char  firstname[80];         //array of characters for the first name
outfile << "Name  " << endl;
infile.get(firstname,80);  //priming the read  reads the first name

while(infile)
{
   infile.get(dummy);     //reads the end of line character into dummy
   outfile << firstname << endl;       //outputs the name
   infile.get(firstname,80);          //reads the next name
}
```

In the above example, `dummy` is used to consume the end of line character.

`input.get(firstname,80);` reads the characters Mary Lou and stops just before reading the `<eol>` end of line character. The `infile.get(dummy)` gets the end of line character into `dummy`.

Another way to do this is with the `ignore` function, which reads characters until it encounters the specific character it has been instructed to look for. The statement `infile.ignore(81,'\n')` skips up to 81 characters stopping if the new line '\n' character is encountered. This newline character IS consumed by this function, and thus there is no need for a dummy character variable.

Example:

```
char  firstname[80];

outfile << "Name  " << endl;
infile.get(firstname,80);

while(infile)
{
   infile.ignore(81,'\n');      // read and consume the end of line character
   outfile << firstname << endl;
   infile.get(firstname,80);
}
```

Still another way to read in lines containing whitespace characters is with the `getline` member function. It needs neither a dummy variable nor the ignore function.

Example:

```
char  firstname[80];
outfile << "Name  " << endl;
infile.getline(firstname,81); //We use a number 1 greater than the total
                              //amount to read.  Lines contain 80 characters
while(infile)
{
   outfile << firstname << endl;
   infile.getline(firstname,81);
}
```

Files Used for Both Input and Output

A file can be used for both input and output. The `fstream` data type, which is used for files that can handle both input and output, must have a **file access flag** as an argument to the open function so that the mode, input or output, can be determined.

Example:

```
#include <fstream.h>                     // needed to use files

fstream test ("grade.dat", ios::out) // test is declared as an
                                     // fstream file first used as output
                                     // ios::out is the file access flag

//   code

test.close();                        // close the file
test.open ("grade.dat", ios::in)     // reopen file as an input file
                                     // ios::in is the file access flag

// Other code goes here

test.close();                        // closes the file
```

Binary Files

So far all the files we have talked about have been text files, files formatted as ASCII text. Even numbers written to a file with the << operator are changed to ASCII text. ASCII is a code that stores every datum (letter of the alphabet, digit, punctuation mark, etc.) as a character with a unique number. Although ASCII text is the default method for storing information in files, we can specify that we want to store data in pure binary format by "opening" a file in binary mode with the `ios::binary` flag. The `write` member function is then used to write binary data to the file. This method is particularly useful for transferring an entire array of data to a file. Binary files are efficient since they store everything as 1's or 0's rather than as text. While efficient it also means that they can only be read by a program and not printed or viewed with a word processor.

Example:

```
fstream test("grade.dat", ios::out| ios::binary);  // This declares and opens
                                                   // the file test as an
                                                   // output binary file

int grade[arraysize] = {98, 88, 78, 77, 67, 66, 56, 78,98, 56}; // creates and
                                                                // initializes
                                                                // an integer
                                                                // array

test.write((char*)grade, sizeof(grade));  // write all values of array to file

test.close();                             // close the file
```

`test.write((char*)grade, sizeof(grade));` in the above example calls the `write` function. The name of the file to be written to is `test`. The first argument is a character pointer pointing to the starting address of memory which is to be written to the file, in this case to the beginning of the `grade` array. The second argument is the size in bytes of the item written to the file. `sizeof` is a function that determines that size.

The following sample program initializes an array and then places those values into a file as binary numbers. The program then adds 10 to each element of the array and prints those values to the screen. Finally the program reads the values from the same file and prints them. These values are the original numbers. Study the program and its comments very carefully.

Sample Program 12.3:

```
#include <fstream.h>        //directive needed to use files
#include <iostream.h>

const arraysize = 10;

void main()
{
   fstream test("grade.dat", ios::out| ios::binary);
   int grade[arraysize] = {98, 88, 78, 77, 67, 66, 56, 78, 98, 56};
   int count;                                   // loop counter
   test.write((char*)grade, sizeof(grade));     // write values of array to file
   test.close();    //close file

   //add 10 to each grade
   cout << "The values of grades with 10 points added" << endl;
   for (count = 0; count < arraysize; count++)
   {
      grade[count] = grade[count] + 10;   //adds 10 to each element of array
      cout << grade[count] << endl;       //write those values to the screen
   }
   test.open("grade.dat", ios::in);        //reopen the file but now as an input
                                           //file

   test.read((char*) grade, sizeof(grade));
```

The above statement reads from the file `test` and places the values found into the `grade` array. As with the `write` function, the first argument is a character pointer even though the array itself is integer. It points to the starting address in memory where the file information is to be transferred.

```
   cout << "The grades as they were read into the file" << endl;
   for (count = 0; count < arraysize; count++)
      cout << grade[count] << endl;            //writing values to the screen

   test.close();

}
```

The output to the screen from this program is as follows:

```
The values of grades with 10 points added
108
98
88
87
77
76
66
88
108
66
The grades as they were read into the file
98
88
78
77
67
66
56
78
98
56
```

Files and Records

Files are often used to store records. A "field" is one piece of information and a "record" is a group of fields that logically belong together in a unit.

Example:

Name	Test1	Test2	Final
Brown	89	97	88
Smith	99	89	97

Each record has four fields: `Name`, `Test1`, `Test2` and `Final`. When records are stored in memory, rather than in files, C++ structures provide a good way to organize and store them.

```
struct Grades
{
     char name[10];
     int test1;
     int test2;
     int final;
};
```

A variable declared to be a `Grades` structure could hold one record.

The `write` function, mentioned in the previous section, can be used to write records to a file.

```
fstream test("score.dat", ios::out|ios::binary);
Grades student;                              //declaring a structure

// code that gets information into the student record
test.write((char *) & student, sizeof(student));
```

Notice that the `test.write` function above used to write a record stored as a `struct` is similar to the `write` function used for an array with one big difference. Notice the inclusion of (&). Why is this necessary here and not with writing an array? This is pass by reference which requires the (&) symbol. Arrays are also passed by reference but with arrays the (&) is not necessary. The following example takes records from the user and stores them into a file.

Sample Program 12.4:

```
#include <fstream.h>     //file directives
#include <iostream.h>

const NAMESIZE = 31;
struct Grades    //defining a structure
{
     char name[NAMESIZE];
     int test1;
     int test2;
     int final;
};

void main ()
{
   fstream tests("score.dat", ios::out| ios::binary); //declare tests as an
                                                      //output binary file.
   Grades student;  //declare student as a record (struct)
   char more;       //used to determine if there is more input
   do
   {
     cout << "Enter the following information" << endl;
     cout << "Student's name: ";
     cin.getline(student.name,NAMESIZE);

     cout << "First test score :";
     cin >> student.test1;
     cin.ignore(); //ignore rest of line

     cout << "Second test score: ";
     cin >> student.test2;
     cin.ignore();

     cout << "Final test score: ";
     cin >> student.final;
     cin.ignore();

     //write this record to the file
     tests.write((char *)&student, sizeof(student));

     cout << "Enter a y if you would like to input more data" << endl;
     cin >> more;
     cin.ignore();
   }  while (more == 'y');
     tests.close();
}
```

PRE-LAB WRITING ASSIGNMENT

Fill-in-the-Blank Questions

1. Default files are ASCII ______________ files.
2. A "record" is specific information of a group of ______________.
3. Files that will be used for both input and output should be declared as ______________ data type.
4. Files that will be used only for input can be declared as ______________ data type.
5. Files that will be used only for output can be declared as ______________ data type.
6. The >> operator ______________ any leading whitespace.
7. ______________ files are files that store data as ones and zeros rather than ASCII text code.
8. The ______________ function "looks ahead" to determine the next data value in an input file.
9. The ______________ function does NOT skip leading whitespace characters.
10. The ______________ function can be used to send an entire record or array to a binary file with one statement.

LESSON 12A

LAB 12.1 Introduction to Files

Retrieve program `files.cpp` from the Lab 12 folder. The code is as follows:

```
// This program uses hours, pay rate, state tax and fed tax to determine gross
// and net pay.

// PLACE YOUR NAME HERE

#include <fstream.h>
#include <iostream.h>
#include <iomanip.h>

int main()
{
    // FILL IN THE CODE TO DECLARE payfile AS AN INPUT FILE
    float gross;
    float net;
    float hours;
    float payRate;
    float stateTax;
    float fedTax;

    cout << setiosflags(ios::fixed) << setprecision(2)
         << setiosflags(ios::showpoint);
    // FILL IN THE CODE TO OPEN payfile AND ATTACH IT TO THE PHYSICAL FILE
    // NAMED payroll.dat
```

continues

```
    // FILL IN CODE TO WRITE A CONDITIONAL STATEMENT TO CHECK IF payfile
    // DOES NOT EXIST.
    {
        cout << "Error opening file. \n";
        cout << "It may not exist where indicated" << endl;
        return 1;
    }

    cout << "Payrate      Hours    Gross Pay        Net Pay" << endl << endl;
    // FILL IN CODE TO PRIME THE READ FOR THE payfile FILE.
    // FILL IN CODE TO WRITE A LOOP CONDITION TO RUN WHILE payfile HAS MORE
    // DATA TO PROCESS.
    {
        payfile >> payRate >> stateTax >> fedTax;
        gross = payRate * hours;
        net = gross - (gross * stateTax) - (gross * fedTax);
        cout << payRate << setw(15) << hours << setw(12) << gross << setw(12)
             << net << endl;
        payfile >> // FILL IN THE CODE TO FINISH THIS WITH THE APPROPRIATE
                   // VARIABLE TO BE INPUT
    }
        payfile.close();
        return 0;
}
```

Exercise 1: Assume that the data file has `hours`, `payrate`, `stateTax` and `fedTax` on one line for each employee. `stateTax` and `fedTax` are given as decimals (5% would be .05). Complete this program by filling in the code (places in bold).

Exercise 2: Run the program. Note: the data file does not exist so you should get the error message:

```
Error opening file.
It may not exist where indicated
```

Exercise 3: Create a data file with the following information:

40 15.00 .05 .12
50 10 .05 .11
60 12.50 .05 .13

Save it in the same folder as the `.cpp` file. What should the data file name be?

Exercise 4: Run the program. Record the output here

____

____

____

____

____

Exercise 5: Change the program so that the output goes to an output file called `pay.out` and run the program. You can use any logical name you wish for the output file.

LAB 12.2 Files as Parameters and Character Data

Retrieve program `Grades.cpp` and the data file `graderoll.dat` from the Lab 12 folder. The code is as follows:

```
#include //  FILL IN THE REST OF THE CODE.  DIRECTIVE FOR FILES
#include <iostream.h>
#include <iomanip.h>

// PLACE YOUR NAME HERE

//  This program reads records from a file.  The file contains the following:
//  a student's name, two test grades and a final exam.   It then prints this
//  information to the screen.

const NAMESIZE = 15;
const MAXRECORDS = 50;
struct Grades                              // define a structure
{
    char name[NAMESIZE];
    int test1;
    int test2;
    int final;

};

typedef Grades gradetype[MAXRECORDS];     // This makes gradetype a data
                                          // data type that holds MAXRECORDS
                                          // Grades structures.

// FIll IN THE CODE FOR THE PROTOTYPE OF THE FUNCTION ReadIt
// WHERE THE FIRST ARGUMENT IS AN INPUT FILE, THE SECOND IS THE
// ARRAY OF RECORDS, AND THE THIRD WILL HOLD THE NUMBER OF RECORDS
// CURRENTLY IN THE ARRAY.

int main()

{
     ifstream indata;
     indata.open("graderoll.dat");
     int numRecord;                    // number of records read in
     gradetype studentRecord;

    if(!indata)
    {
        cout << "Error opening file. \n";
        cout << "It may not exist where indicated" << endl;
        return 1;
    }

    // FILL IN THE CODE TO CALL THE FUNCTION ReadIt.
```

continues

```
    // output the information
    for (int count = 0; count < numRecord; count++)
    {
       cout << studentRecord[count].name << setw(10) << studentRecord[count].test1
            << setw(10) << studentRecord[count].test2;
       cout << setw(10) << studentRecord[count].final << endl;
    }

    return 0;
}

// This function reads in records into an array of records from an input file and
// keeps track of the total number of records

void ReadIt(// FILL IN THE CODE FOR THE FORMAL PARAMETERS AND THEIR DATA TYPES.
            // indata, graderec and total are the formal parameters
            // total is passed by reference

{
    total = 0;

    indata.get(graderec[total].name, NAMESIZE);
    while (indata)
    {
      // FILL IN THE CODE TO READ test1
      // FILL IN THE CODE TO READ test2
      // FILL IN THE CODE TO READ final

      total++;      // add one to total

      // FILL IN THE CODE TO CONSUME THE END OF LINE
      // FILL IN THE CODE TO READ name

  }

}
```

Exercise 1: Complete the program by filling in the code. (Areas in bold) This problem requires that you study very carefully the code already written and the data file to prepare you to complete the program. Notice that in the data file the names occupy no more than 15 characters. Why?

Exercise 2: (Optional) Add another field to the record called `letter` which is a character that holds the letter grade of the student, which is based on the average of the grades as follows: test1 and test2 are each worth 30% of the grade while final is worth 40% of the grade. The letter grade is based on a 10 point spread. The code will have to be expanded to find the average.

90–100	A
80–89	B
70–79	C
60–69	D
0–59	F

LESSON 12B

LAB 12.3 Binary Files and the `write` Function

Retrieve program `budget.cpp` from the Lab 12 folder. The code is as follows:

```
// This program reads in from the keyboard a record of financial information
// consisting of a person's name, income, rent, food, utilities and
// miscellaneous expenses.  It then determines the net money
// (income minus all expenses)and places that information in a record
// which is then written to an output file.

// PLACE YOUR NAME HERE

#include <fstream.h>
#include <iostream.h>
#include <iomanip.h>

const NAMESIZE = 15;
struct  budget  //define a structure to hold name and financial information
{
   char name[NAMESIZE];
   float income;        // person's monthly income
   float rent;          // person's monthly rent
   float food;          // person's monthly food bill
   float utilities;     // person's monthly utility bill
   float miscell;       // person's other bills
   float net;           // person's net money after bills are paid

};

void main()

{
   fstream indata;
   ofstream outdata;                                     // output file of
                                                         // student
   indata.open("income.dat", ios::out | ios::binary);   // open file as binary
                                                         // output
   outdata.open("person.out");                           // output file that
                                                         // we'll write student
                                                         // information to
   outdata << setiosflags(ios::left)                     // left indicates left
           << setiosflags(ios::fixed) << setprecision(2);   // justified for fields

   budget person;    //declares person to be a record

   cout << "Enter the following information" << endl;
   cout << "Person's name: ";
   cin.getline(person.name, NAMESIZE);
   cout << "Income :";
```

continues

```
    cin >> person.income;

    // FILL IN CODE TO READ IN THE REST OF THE FIELDS:
    // rent, food, utilities AND miscell TO THE person RECORD

    // find the net field
    person.net = // FILL IN CODE TO DETERMINE NET INCOME (income - expenses)

    // write this record to the file
    // Fill IN CODE TO WRITE THE RECORD TO THE FILE indata (one instruction)

    indata.close();

    // FILL IN THE CODE TO REOPEN THE indata FILE, NOW AS AN INPUT FILE.
    // FILL IN THE CODE TO READ THE RECORD FROM indata AND PLACE IT IN THE
    // person RECORD (one instruction)

    // write information to output file
    outdata << setw(20) << "Name" << setw(10) << "Income" << setw(10) << "Rent"
            << setw(10) << "Food" << setw(15) << "Utilities" << setw(15)
            << "Miscellaneous" << setw(10) << "Net Money" << endl << endl;

    // FILL IN CODE TO WRITE INDIVIDUAL FIELD INFORMATION OF THE RECORD TO
    // THE outdata FILE.(several instructions)

}
```

Exercise 1: This program reads in a record with fields `name`, `income`, `rent`, `food`, `utilities`, and `miscellaneous` from the keyboard. The program computes the net (income minus the other files) and stores this in the `net` field. The entire record is then written to a binary file (`indata`). This file is then closed and reopened as an output file where the record is stored. Fill in the code as indicated by the comments in bold.

Sample Run:

```
Enter the following information
Person's Name: Billy Berry
Income: 2500
Rent: 700
Food: 600
Utilities: 400
Miscellaneous: 500
```

The program would write the following text lines to the output file `student.out`.

```
Name          Income     Rent      Food      Utilities  Miscellaneous  Net Money

Billy Berry   2500.00    700.00    600.00    400.00     500.00         300.00
```

Exercise 2: Alter the program to include more than one record as input. Make arrays of records.

Sample Run:

```
Enter the following information
Person's Name: Billy Berry
Income: 2500
Rent: 700
Food: 600
Utilities: 400
Miscellaneous: 500

Enter a Y if you would like to input more data
Y

Enter the following information
Person's Name: Terry Bounds
Income: 3000
Rent: 750
Food: 650
Utilities: 300
Miscellaneous: 400

Enter a Y if you would like to input more data
n
That's all the information
```

The output file student.out would then have the following lines of text written to it.

Name	Income	Rent	Food	Utilities	Miscellaneous	Net Money
Billy Berry	2500.00	700.00	600.00	400.00	500.00	300.00
Terry Bounds	3000.00	750.00	650.00	300.00	400.00	900.00

LAB 12.4 Student Generated Code Assignments

Option 1: Write a program that will: 1) read an array of records from the keyboard, 2) store this information to a binary file, 3) read from the binary file back to the array of records, 4) store this information to a textfile. Left justify the information for each field. Each record will consist of the following fields:

first name	15 characters
last name	15 characters
street address	30 characters
city	20 characters
state	5 characters
zip	long integer

You may assume a maximum of 20 records.

This assignment is very similar to the program found in Lab 12.3.

Sample Run:

```
Enter the following information
Person's First Name: Billy
Person's Last Name: Berry
Street: 205 Main Street
City: Cleveland
State: Tx
Zip: 45679

Enter a Y if you would like to input more data
Y

Enter the following information
Person's First Name: Sally
Person's Last Name: Connely
Street: 348 Wiley Lane
City: San Francisco
State: Md
Zip: 54789

Enter a Y if you would like to input more data
n
That's all the information
```

The output file contains the following:

```
First Name  Last Name  Street           City           State  Zip Code

Billy       Berry      205 Main Street  Cleveland      Tx     45679
Sally       Connely    348 Wiley Lane   San Francisco  Md     54789
```

Option 2: Write a program that will read the radii of circles. This is an array of records where each record will have the radius of the circle read from the keyboard and the diameter and area of the circle will be calculated by the program. This information (radius, diameter and area) is stored in a binary file. The information in the binary file is then read back into the records and stored in a text output file. Left justify the information for each field. Each record will consist of the following fields:

radius	float
diameter	float
area	float

You may assume a maximum of 20 records. You may want to include the `math.h` library.

You need to know the formulas for finding the area and circumference of a circle.

This assignment is very similar to the program in Lab 12.3.

Sample Run:

```
Enter the following information:
Radius of circle: 5
Enter a Y if you would like to input more data
y
```

```
Enter the following information:
Radius of circle: 4
Enter a Y if you would like to input more data
y

Enter the following information:
Radius of circle: 7
Enter a Y if you would like to input more data
n

That's all the information.
```

The output file contains the following:

```
Radius   Area    Circumference

5.00      78.54   31.42
4.00      50.27   25.13
7.00     153.94   43.98
```

Option 3: Bring in the file `employee.in` from Lab 12 folder. Write a program that will read records from this file and store them in a binary file. That file will then be used as input to create an output file of the information. The data file contains employee information consisting of name, social security, department id, years employed. In addition to displaying the information of each record the program will also calculate the average salary and years employed of all the records. This additional information is stored in that same output file.

Sample Data File:

Bill Tarpon	182460678	789	8	30600
Fred Caldron	456905434	789	10	40700
Sally Bender	203932239	790	8	50000
David Kemp	568903493	790	9	60000

The output file should look like this:

```
Name          Social Security  Department ID  Years Employed  Salary

Bill Tarpon   182460678        789                  8          30600.00
Fred Caldron  456905434        789                 10          40700.00
Sally Bender  203932239        790                  8          50000.00
David Kemp    568903493        790                  9          60000.00

The average number of years employed is 8

The average salary is $45325.00
```

LESSON SET

13.1 Introduction to Classes

PURPOSE

1. To introduce object-oriented programming
2. To introduce the concept of classes

PROCEDURE

1. Students should read Chapter 13 of the text.
2. Students should read the Pre-lab Reading Assignment before coming to lab.
3. Students should complete the Pre-lab Writing Assignment before coming to lab.
4. In the lab, students should complete labs assigned to them by the instructor.

Contents	Pre-requisites	Approximate completion time	Page number	Check when done
Pre-lab Reading Assignment	Chapter 13 of text	20 min.	200	
Pre-lab Writing Assignment	Pre-lab reading	10 min.	207	
LESSON 13.1A				
Lab 13.1 Money Exchange	Basic understanding of structures and classes	35 min.	207	
Lab 13.2 Square as a Class	Completion of Pre-lab Reading Assignment	15 min.	209	
LESSON 13.1B				
Lab 13.3 Arrays as Data Members of Classes	Understanding of private data members of classes and files	20 min.	210	
Lab 13.4 Student Generated Code Assignments	Completion of all the previous labs	30 min.	212	

PRE-LAB READING ASSIGNMENT

Introduction to Object-Oriented Programming

Up until now, we have been using the procedural programming method for writing all our programs. A procedural program has data stored in a collection of variables (or structures) and has a set of functions that perform certain operations. The functions and data are treated as separate entities. Although operational, this method has some serious drawbacks when applied to very large real-world situations. Even though this process is modularized (broken into several functions), in a large complex program the number of functions can become overwhelming and difficult to modify or extend. This can create a level of complexity that is difficult to understand.

Object-Oriented Programming (OOP) mimics real world applications by introducing **classes** which act as a prototype for **objects**. Objects are similar to nouns that can simulate persons, places or things that exist in the real world. OOP enhances code reuse ability (use of existing code or classes) so time is not used on "reinventing the wheel."

Classes and objects are often confused with one another; however, there is a subtle but important difference best explained by the following example. A plaster of Paris mold consists of the design of a particular figurine. When the plaster is poured into the mold and hardened, we have the creation of the figurine itself. A class is analogous to the mold, for it holds the definition of an object. The object is analogous to the figurine, for it is an **instance** of the class. Classes and structures are very similar in their construction. The main difference is that classes may encapsulate functions with data declarations. Object-oriented programming is not learned in one lesson. This lab gives a brief introduction into this most important concept of programming.

A **class** is a prototype (template) for a set of objects. An object can be described as a single instance of a class in much the same way that a variable is an instance of a particular data type. A class is the best way to describe an **abstract data type (ADT)** which is described as a user defined data type that combines a collection of variables and operations. A class consists of a name (its identity), its member data (what it is), and its member functions (what it does).[1] For example, a circle, in order to be drawn, must have a center (an x and a y coordinate) and a radius. In practical terms we could describe this as its member data (`radius`, `center_x`, `center_y`). We can also describe a set of member functions that can give values to the data and perform certain actions with the circle. These actions can consist of giving values for the center and radius and finding the circumference and area of the circle. A class thus combines data and functions under one package. An object should protect its important data by declaring it as private and then provide a public interface to access that data. C++ uses the following code to define the `Circles` class:

[1] In other object-oriented languages member functions are called methods and member data are called attributes.

```
class Circles      // Circles is the name of the class (its identity).
                   // The following is the declaration of the Circles class

{

public:
  // The following is labeled as public since they can be used by any object.
  // Usually member functions are public while member data are private.
  // Functions describe what the class does.

  void SetCenter(int x, int y);
    // This member function inputs the center coordinates of the circle object
    // that calls it.

  void SetRadius(float r);
    // This member function inputs the radius of the circle object that calls
    // it.

  double AreaOf();
    // This member function finds the area of the circle object that calls it.

  double Circumference();
    // This member function finds the circumference of the circle object that
    // calls it.

  void CircleStats();
    // This member function outputs the radius and center of the circle object
    // that calls it.

private:
  // Member data are usually defined as private which means that they can ONLY
  // be accessed by the object.
  // Member data describes what the class is made of

  float  radius;
  float  center_x;
  float  center_y;
};                        // Notice the braces and the semicolon at the end.
```

It is not necessary for someone to understand how a television remote control works in order to use the remote to change the stations or the volume. The user of the remote could be called a **client** that only knows how to use the remote to accomplish a certain task. The details of how the remote control performs the task are not necessary for the user to use the remote. Likewise, an automobile is a complex mechanical machine with a simple interface that allows users without any (or very little) mechanical knowledge to start, drive and use it for a variety of functions. Drivers do not need to know what goes on under the hood. In the same way, a program that uses `Circles` does not need to know the details of how its member functions perform their operations. The use of an object (an instance of a class) is thus separated into two parts: the **interface** (client file) which calls the functions and the **implementation** which contains the details of how the functions accomplish what they do.

An object not only combines data and functions but also restricts other parts of the program from accessing member data and the inner workings of member functions. The concept of having programs or users access only certain parts of an object is known as **data hiding**. The fact that the internal data and inner workings can be hidden from users makes the object more accessible to a greater number of programs.

Just like an automobile or a remote control, commercial software is a complex entity developed by many individuals. OOP (Object-Oriented Programming) allows programmers to create objects with hidden complex logic that have a simple **interface** which is easily understood and used. This allows more sophisticated programs to be developed. Interfacing is a major concern for software developers.

User of an object → Public Interface → | Private Internal Data (center_x, center_y, radius) Implementation of the member functions |

Types of Objects

Objects are either general purpose or application-specific. General purpose objects are designed to create a specific data type such as currency or date. They are also designed to perform common tasks such as input verification and graphical output. Application-specific objects are created as a specific limited operation for some organization or task. A student class, for example, may be created for an educational institution.

Implementations of Classes in C++

The class definition is usually placed in the global section of a program or in a special file (called a **header** file). As noted, the class definition acts very much like a prototype or data type for an object. An object is declared much like a variable except that it uses the class name as the data type. This declaration creates an **instance** (actual occurrence) of the class. Implementation of the member functions of a class are given either after the `main` function of the program or in a separate file called the **implementation** file. Use of the object is usually in the `main` function, other specialized functions, or in a separate program file called the **client** file.[2]

Use of Objects

`Circles` described above is a class (prototype) and not an object (an actual instance of the class). Objects are declared in the client file, `main`, or other functions just as variables are declared:

```
Circles sphere1,sphere2;
```

`sphere1` and `sphere2` are objects of class `Circles`.

`sphere1` has it's own `radius`, `center_x` and `center_y` that are possibly different from the `radius`, `center_x` and `center_y` of `sphere2`.

---

[2] More will be given on header, implementation and client files later in the lesson.

To access a particular member function, an object must list it's own name and the name of the function separated from each other by a period.

Examples:

```
sphere1.SetCenter(9,10);   // Notice we have associated a particular object
                           // with the call to the SetCenter function.
sphere1.SetRadius(20);

cout << "The area of the circle is " << sphere1.AreaOf() << endl;
```

Since `AreaOf` must have the radius of the object before it can do the calculation, an object must call `SetRadius` before it calls `AreaOf`. The user must remember to initialize a radius (by calling `SetRadius`). It is not good programming practice to assume that a user will do the necessary initialization. Constructors (discussed in Lesson Set 13.2) solves this problem.

The following is a complete `main` function (or client file) that declares and uses objects which call member functions.

```
void main()
{
   Circles  sphere1;      // sphere1 is declared as an object of Circles class
   float area;            // area contains the area of the circle
   float circum;          // circum contains the circumference of the circle


   sphere1.SetCenter(9,10);
   sphere1.SetRadius(20);

   cout << "The area of the circle is " << sphere1.AreaOf() << endl;
   cout << "The circumference of the circle is "
        << sphere1.Circumference() << endl;

   sphere1.CircleStats();

}
```

Implementation of Member Functions

As previously noted, the implementation of the member function can be hidden from the users (clients) of the objects. However, they must be implemented by someone, somewhere. The following shows the implementation of the `Circles` member functions.

```
void Circles::SetCenter(int x, int y)
   // This procedure receives the coordinates of the center of the circle
   // from the user and places them in the appropriate member data.
{
   center_x = x;
   center_y = y;
}
```

continues

```
void Circles::SetRadius(float r)
   // This procedure will take the radius of the circle from the user and
   // place it in the appropriate member data
{
   radius = r;
}

double Circles::AreaOf()
   // This function will return the area of the circle object that calls it.
{
   return PI * radius * radius;
}

double Circles::Circumference()
   // This function will return the circumference of the circle object that
   // calls it.
{
   return 2 * PI * radius;
}

void Circles::CircleStats()
   // This procedure prints out the radius and center coordinates of the
   // circle object that calls it.
{
   cout << "The radius of the circle is " << radius << endl;
   cout << "The center of the circle is " << center_x
        << "    " << center_y << endl;
}
```

Usually classes are defined in a header file while member functions are stored in an implementation file and objects are declared and used in a client file. These files are often bound together in a project. Various development environments have different means of creating and storing related files in a project.[3] All could be located in three different sections of the same file.

Complete Program

The following code shows the class definition, member function (method) implementations and use (client) of the `Circles` class:

```
#include <iostream.h>

//________________________________________________________________
// Class definition and constant declaration section   (header file)

class Circles        // Circles is the name of the class (its identity).
                     // The following is the declaration of the class of Circle
                     // objects.
```

[3] Check with the instructor for the environment method used in this class.

```
{
public:
  // The following is labeled as public since they can be used by any object.
  // Usually member functions are public while member data are private.
  // Functions describe what the class does.

  void SetCenter(int x, int y);
    // This member function inputs the center coordinates of the circle object
    // that calls it.

  void SetRadius(float r);
    // This member function inputs the radius of the circle object that calls
    // it.

  double AreaOf();
    // This member function finds the area of the circle object that calls it.

  double Circumference();
    // This member function finds the circumference of the circle object that
    // calls it.

  void CircleStats();
    // This member function outputs the radius and center of the circle object
    // that calls it.

private:
  // Member data are usually defined as private which means that they can ONLY
  // be accessed by the object.
  // Member data describes what the class is made of

  float  radius;
  float  center_x;
  float  center_y;
};                       // Notice the braces and the semicolon at the end.

const double PI = 3.14;

//____________________________________________________________________________
// Client file

void main()

{
   Circles  sphere1;       // sphere1 is declared as an object of Circles class

   sphere1.SetCenter(9,10);
   sphere1.SetRadius(20);

   cout << "The area of the circle is " << sphere1.AreaOf() << endl;
   cout << "The circumference of the circle is "
        << sphere1.Circumference() << endl;

   sphere1.CircleStats(); // defined in the next section

}
```

continues

```
//______________________________________________________________________
// Implementation section     Member function implementation

void Circles::SetCenter(int x, int y)
   // This procedure receives the coordinates of the center of the circle
   // and places them in the appropriate member data.
{
   center_x = x;
   center_y = y;
}

void Circles::SetRadius(float r)
   // This procedure will take the radius of the circle and
   // place it in the appropriate member data
{
   radius = r;
}

double Circles::AreaOf()
   // This function will return the area of the circle object that calls it.
{
   return PI * radius * radius;
}

double Circles::Circumference()
   // This function will return the circumference of the circle object that
   // calls it.
{
   return 2 * PI * radius;
}

void Circles::CircleStats()
   // This procedure prints out the radius and center coordinates of the
   // circle object that calls it.
{
   cout << "The radius of the circle is " << radius << endl;
   cout << "The center of the circle is " << center_x
        << "   " << center_y << endl;
}
```

Inline Member Functions

Sometimes the implementation of member functions is so simple that they can be defined inside a class declaration. Such functions are called inline member functions. In the above `Circles` class, `AreaOf` and `Circumference` are so simple that they could be defined in the class definition as follows:

```
const double PI = 3.14;
class Circles

private:

     float  radius;
     float  center_x;
     float  center_y;
```

```
public:

     void SetCenter(int x, int y);
     void SetRadius(float r);
     double AreaOf()    { return PI * radius * radius; }
     double Circumference() { return 2 * PI * radius; }
     void CircleStats();
};
```

PRE-LAB WRITING ASSIGNMENT

Fill-in-the-Blank Questions

1. A(n) _______________ is an instance of a class.
2. Member functions are called _______________ in other object-oriented languages.
3. A header file contains the _______________ definitions.
4. Member functions are defined in an _______________ file.
5. To access a particular member function, an object must list it's own name and the name of the function separated from each other by a _______________.
6. _______________ programming combines both data and functions that operate on the data together in one package.
7. _______________ member data can only be accessed by its corresponding object.
8. A _______________ is the best way to describe an abstract data type (ADT).
9. _______________ hiding allows programmers to create objects with hidden complex logic, that have a simple interface that is easily understood and used.
10. An _______________ member function has its implementation given in the class definition.

LESSON 13.1A

LAB 13.1 Money Exchange

Retrieve program `moneychange.cpp` from the Lab 13.1 folder. The code is as follows:

```
// This program is the money change program.  It will convert dollars
// to lire and francs.

// PLACE YOUR NAME HERE

#include <iostream.h>

const float DOLLAR_LIRE = 2048;          // Lire conversion per dollar
const float DOLLAR_FRANC = 6.937;        // Franc conversion per dollar

class Money
{
```

continues

```
public:
    void GetMoney(long, long);
    float Lire() const;       // The word const after a member function
                              // indicates that the function will not alter
                              // member data.  This function returns Italian
                              // currency of current dollar and cent amount

    float Franc()const;       // This function returns French currency of
                              // current dollar and cent

    long Dollar() const;      // This function returns dollar amount

    long Cents() const;       // This function returns cents amount

private:

    long dollars;
    long cents;
};

// client file

void main()
{

    // Fill in the code to declare an object called money1 of type Money class
    // Fill in the code to declare an object called money2 of type Money class

    // Fill in the code to call GetMoney so that object money1 has 10 dollars
    // 59 cents

    // Fill in the code to call GetMoney so that object money2 has 20 dollars
    // 70 cents

    cout << "The value of " << money1.Dollar() << " dollars and "
         << money1.Cents() << " cents in Lire is " << money1.Lire() << endl;

    // Study the above cout statement.  Fill in the code for a similar
    // statement that will return the value of money1 in Francs

    // Fill in the code for two cout statements that will do the same thing
    // for money2.

}

//Implementation file

void Money::GetMoney(long dollaramount, long centamount)

{
    dollars = dollaramount + centamount / 100;
    cents = centamount % 100;
}

float Money::Lire() const
{
```

```
    // Fill in the code that will return the lire value of dollars and cents
    // of the object that calls it.

}

float Money::Franc() const

{

    // Fill in the code that will return the franc value of dollars and cents
    // of the object that calls it.
}

long Money::Dollar() const
{
    return dollars;
}

long Money::Cents() const
{
    return cents;
}
```

Exercise 1: Fill in the code for the client and implementation so that the following output is obtained:

The value of 10 dollars and 59 cents in Lire is 21688.3
The value of 10 dollars and 59 cents in Francs is 73.4628
The value of 20 dollars and 70 cents in Lire is 42393.6
The value of 20 dollars and 70 cents in Francs is 143.596

Exercise 2: Add the appropriate code to convert the money to Pesos (1 dollar = 9.815 Pesos) so that the following two output statements will be added.

```
The value of 10 dollars and 59 cents in Pesos is 103.941
The value of 20 dollars and 70 cents in Pesos is 203.17
```

LAB 13.2 Squares as a Class

Retrieve program `square.cpp` from the Lab 13.1 folder. The code is as follows:

```
// This program defines Square as a class and uses member functions to find the
// perimeter and area of the square

// PLACE YOUR NAME HERE

#include <iostream.h>

// Fill in the code to define the class called Square. This can be determined
// by looking at the implementation section.

//___________________________________________________________________________
// Client file

void main()
{
```

continues

```
    Square  box;            // box is declared as an object of Square class
    float   size;           // size contains the length of a side of the square

    // FILL IN THE CLIENT CODE THAT WILL ASK USER FOR THE LENGTH OF THE SIDE
    // OF THE SQUARE.  (This is stored in size)

    // FILL IN THE CODE THAT CALLS SetSide.

    // FILL IN THE CODE THAT WILL RETURN THE AREA FROM A CALL TO A FUNCTION
    // AND PRINT OUT THE AREA TO THE SCREEN

    // FILL IN THE CODE THAT WILL RETURN THE PERIMETER FROM A CALL TO A
    // FUNCTION AND PRINT OUT THAT VALUE TO THE SCREEN

}

//___________________________________________________________________________
//Implementation section     Member function implementation

void Square::SetSide(float length)
    // This procedure will take the length of a side and
    // place it in the appropriate member data
{
    side = length;
}

float Square::AreaOf()
{
    return side * side;
}

float Square::Perimeter()
{
    return 4 * side;
}
```

Exercise 1: This program asks you to fill in the class definition and client code based on the implementation of the member functions. Fill in the code so that the following input and output will be generated:

```
Please input the size of the side of the square
8
The area of the square is 64
The perimeter of the square is 32
```

LESSON 13.1B

LAB 13.3 Arrays as Data Members of Classes

Retrieve program `floatarray.cpp` and `temperatures.txt` from the Lab 13.1 folder. The code is as follows:

```
// This program reads floating point data from a data file and places those
// values into the private data member called values (a floating point array)
// of the Floatlist class. Those values are then printed to the screen.
```

```
// The input is done by a member function called GetList, the output
// is done by a member function called PrintList.  The amount of data read in
// is stored is the private data member called length.  The member function
// GetList is called first so that length can be initialized to zero.

// PLACE YOUR NAME HERE

#include <iostream.h>
#include <fstream.h>
#include <iomanip.h>

const int MAX_LENGTH = 50; // MAX_LENGTH contains the maximum length of our list
class FloatList            // Defines a class that contains an array of floating
                           // point numbers
{
public:
    void GetList(ifstream&);    // Member function that gets data from a file
    void PrintList() const;     // Member function that prints data from that
                                // file to the screen.
    void InitializeList();      // Member function that sets length to 0.
                                // This must be called first!!

private:
    int length;                 // Holds the number of elements in the array
    float values[MAX_LENGTH];   // The array of values

};

void main()
{
    ifstream tempData;    // Declares a data file

    // Fill in the code to declare an object called list of the class FloatList

    cout << setiosflags(ios::fixed) << setiosflags(ios::showpoint);
    cout << setprecision(2);

    tempData.open("temperatures.txt");

    // Fill in the code that calls the InitializeList function.
    // Fill in the code that calls the GetList function.
    // Fill in the code that calls the PrintList function.

}

void FloatList::InitializeList()
{
    // Fill in the code to complete this function
    // This function sets the private data member length to 0
}

// Fill in the entire code for the  GetList function
// The GetList function reads the data values from a data file
// into the the values array of the class FloatList

// Fill in the entire code for the  PrintList function
// The PrintList function prints to the screen the data in
// the values array of the class FloatList
```

This program has an array of floating point numbers as a private data member of a class. The data file contains floating point temperatures which are read by a member function of the class and stored in the array.

Exercise 1: Why does member function `PrintList` have a `const` after its name but `GetList` and `InitializeList` do not?

Exercise 2: Fill in the code so that the program reads in the data from the temperature file and prints them to the screen with the following output:

```
78.90
87.40
60.80
70.40
75.60
```

Exercise 3: Add code (member function, call and function implementation) to print the average of the numbers to the screen so that the output will look like the output from Exercise 2 plus the following:

```
The average temperature is 74.62
```

LAB 13.4 Student Generated Code Assignments

Give a C++ class declaration called `SavingsAccount` for the following information:

Operations (Member Functions)

1. Open account (with an initial deposit). This is called to put initial values in dollars and cents.
2. Make a deposit. A function that will add value to dollars and cents
3. Make a withdrawal. A function that will subtract values from dollars and cents
4. Show current balance. A function that will print dollars and cents.

Data (Member Data)

1. `dollars`
2. `cents`

Give the implementation code for all the member functions.

NOTE: You must perform normalization on cents. This means that if `cents` is 100 or more, it must increment `dollars` by the appropriate amount. Example: if cents is 234, then dollars must be increased by 2 and cents reduced to 34. Study the `moneyexchange` program in the Pre-lab Reading Assignment.

Write code that will create an object called `bank1`. The code will then initially place \$200.50 in the account. The code will deposit \$40.50 and then withdraw \$100.98. It will print out the final value of dollars and cents.

The following output should be produced:

```
Dollars = 40 cents = 2.
```

Part 2: Change the program to allow the user to input the initial values, deposit and withdrawal.

Example:

```
Please input the initial dollars
402

Please input the initial cents
78

Would you like to make a deposit? Y or y for yes
y
Please input the dollars to be deposited
35
Please input the cents to be deposited
67
Would you like to make a deposit? Y or y for yes
y
Please input the dollars to be deposited
35
Please input the cents to be deposited
67
Would you like to make a deposit? Y or y for yes
n
Would you like to make a withdrawal Y or y for yes
y
Please input the dollars to be withdrawn
28
Please input the cents to be withdrawn
08
Would you like to make a withdrawal Y or y for yes
y
Please input the dollars to be withdrawn
75
Please input the cents to be withdrawn
78
Would you like to make a withdrawal Y or y for yes
n

Dollars = 370 Cents = 26
```

LESSON SET

13.2 Constructors and Destructors

PURPOSE

1. To introduce the concept of constructors and destructors
2. To introduce arrays of objects

PROCEDURE

1. Students should complete the reading of Chapter 13 of the text.
2. Students should read the Pre-lab Reading Assignment before coming to lab.
3. Students should complete the Pre-lab Writing Assignment before coming to lab.
4. Students should complete labs assigned to them by the instructor.

Contents	Pre-requisites	Approximate completion time	Page number	Check when done
Pre-lab Reading Assignment	Chapter 13 of text	20 min.	216	
Pre-lab Writing Assignment	Pre-lab reading	10 min.	220	
LESSON 13.2A				
Lab 13.5 Constructors Used in the `Circles` Class	Understanding of constructors, destructors and overloading	35 min.	220	
Lab 13.6 Constructors in Money Exchange	Understanding of constructors	25 min.	223	
LESSON 13.2B				
Lab 13.7 Array of Objects	Understanding of arrays of objects and files	20 min.	225	
Lab 13.8 Student Generated Code Assignment	Completion of Above Labs	35 min.	227	

PRE-LAB READING ASSIGNMENT

Introduction to Constructors

We begin this lab by reexamining the `Circles` class of the last lesson set.

```
class Circles
{
public:

   void SetCenter(int x, int y);
     // This member function inputs the center coordinates of the circle
     // object that calls it.

   void SetRadius(float r);
     // This member function inputs the radius of the circle object that calls
     // it.

   double AreaOf();
     // This member function finds the area of the circle object that calls it

   double Circumference();
     // This member function finds the circumference of the circle object that
     // calls it.

   void CircleStats();
     // This member function outputs the radius and center of the circle
     // object that calls it.

private:

   float  radius;
   float  center_x;
   float  center_y;
};
```

Of the five member functions: `SetCenter`, `SetRadius`, `AreaOf`, `Circumference` and `CircleStats`, one of these must be called before some of the other functions can be called. Can you guess which one? By its very definition, a circle must have a radius. Once an object of type `Circles` is declared the user **must** call `SetRadius`. What happens if the user forgets? Suppose the user tries calling the function `AreaOf` without first calling `SetRadius`. The program will try to find the area of a circle that has no radius. The creator of the class should never rely on the user to initialize essential data.

C++ provides a mechanism, called a **constructor,** to guarantee the initialization of an object. A constructor is a member function that is IMPLICITLY invoked whenever a class instance is created (whenever an object is declared). A constructor is unique from other member functions in two ways:

1. It has the same name as the class itself
2. It does not have a data type (or the word `void`) in front of it. It's only purpose is to initialize an object's private data.

The following shows the `Circles` class using two constructors that replace the `setRadius` function.

```
class Circles
{
public:

     void SetCenter(int x, int y);
     double AreaOf();
     double Circumference();
     void CircleStats();
     Circles(float r); //Constructor allowing a user to input the radius
     Circles();        //Constructor using a default value of a radius

private:

     float  radius;
     float  center_x;
     float  center_y;
};
```

The class includes two constructors, differentiated by their parameter lists. Recall from Lesson Set 6.2 that two or more functions could have the same name as long as their parameters differ in quantity or data type. The parameterless constructor (the second one in the above example) is the **default constructor.** Like all member functions, constructors are defined in the implementation file (or function definition section of a program.)

Constructor Definitions

The function definitions of the two constructors in the above example are as follows:

```
Circles::Circles(float r)
{
     radius = r;
}

Circles::Circles()
{
     radius = 1;
}
```

The first constructor allows the user to input a value for `radius` at the same time that an object is declared (shown later in the lab). The second constructor (the default constructor) will set `radius` to 1 whenever the object is declared. With the use of these constructors, every object of class `Circles` will have a value for `radius`.

Invoking a Constructor

Although a constructor is a member function, it is never invoked using the dot notation. It is invoked when an object (an instance of the class) is declared.

Example:

```
Circles sphere1(8);
Circles sphere2;
```

In the above example, `sphere1` is an object of `Circles` class that has its radius set to 8. Since it has a parameter, it activates the constructor that has a parameter. The object `sphere2` is declared with a radius of 1. It activates the default constructor which initializes the private data `radius` to 1.

Destructors

A **destructor** is a member function that is automatically called to destroy an object. Just like constructors, a destructor has the same name as the class; however, it is preceded by a tilde (~). Destructors are important in that they free up memory used by the object when the object is no longer needed. The destructor is automatically called when an object of the class of that destructor no longer exists. This occurs when the function, where the object is declared, ends.

The following example shows how constructors and destructors operate.

Example:

```
#include <iostream.h>

class Demo
{
public:
   Demo();        // Default constructor
   ~Demo();       // Destructor
};

void main()
{
   Demo demoObj; // demoObject is declared which invokes the default
                 // constructor that prints the message "The constructor has
                 // been invoked"

   cout << "The program is now running" << endl;

}

// Now that the main program is over, the object demoObj is no longer active.
// The destructor is invoked and the message "The destructor has been
// invoked" is printed.

Demo::Demo()
{
   cout << "The constructor has been invoked" << endl;
}

Demo::~Demo()
{
   cout << "The destructor has been invoked" << endl;
}
```

What order do you think the three `cout` statements will be executed?

A class can have only **one** default constructor and **one** destructor.

Arrays of Objects

Arrays can also contain objects of a class. In our `Circles` example, we could have an array of `Circles` objects.

Example:

```
Circles  spheres[4];  // spheres is declared as an array of
                      // objects of Circles class
```

This statement makes an array of 4 elements each consisting of an object of the `Circles` class.

Since this class has a default constructor, the default values are assigned to each element (object) of the array. The radius for each of the objects in the `spheres` array is equal to 1 since that is the default value assigned by the default constructor.

Look at the following code:

```
#include <iostream.h>

class Circles
{
public:
   void SetCenter(int x, int y);
   double AreaOf();
   double Circumference();
   void CircleStats();
   Circles (float r);      // Constructor
   Circles();              // Default constructor
   ~Circles();             // Destructor
private:

   float  radius;
   int    center_x;
   int    center_y;
};

const double PI = 3.14;
const int NUMOFOBJECTS = 4;

//____________________________________________________________
//Client file

void main()

{
   Circles  spheres[NUMOFOBJECTS];  // Spheres is declared as an array of
                                    // objects of Circles class

   for (int pos = 0; pos < NUMOFOBJECTS; pos++)
```

continues

```
        {

            cout << "Stats for circle number " << pos+1 << endl << endl;
            spheres[pos].CircleStats();

            cout << "The area of the circle is " << spheres[pos].AreaOf() << endl;
            cout << "The circumference of the circle is"
                 << spheres[pos].Circumference() << endl << endl;
        }

}
```

PRE-LAB WRITING ASSIGNMENT

Fill-in-the-Blank-Questions

1. A ______________ is used in C++ to guarantee the initialization of a class instance.
2. A constructor has the ______________ name as the class itself.
3. A ______________ constructor has no parameters.
4. Two or more constructors can be declared in a class as long as their ______________ differ in quantity or data type.
5. A ______________ is a member function that is automatically called to destroy an object.
6. A ______________ precedes the destructor name which is the same as the class in which it is defined.
7. A constructor is invoked when an object is ______________.
8. In an array of objects the default constructor is applied to ______________ object in the array.
9. Neither the word ______________ nor a data type precedes the definition of a constructor in a class.
10. A constructor is a member function that is ______________ invoked whenever a class instance is created.

LESSON 13.2A

LAB 13.5 Constructors Used in the `Circles` Class

Retrieve program `circle.cpp` from the Lab 13.2 folder. This is similar to the program given in the Pre-lab Reading Assignment. The code is as follows:

```
//______________________________________________________________
// This program defines a class for a circle that will have
// member functions that set the center, find the area, find
// the circumference and display these attributes.
// The program as written does not allow the user to input data, but
// rather has the values of the circles (spheres in the program)
// initialized  at declaration or set by a function.

//class definition section   (header file)
```

```
// PLACE YOUR NAME HERE

#include <iostream.h>

class Circles
{
public:
    void SetCenter(int x, int y);
    double AreaOf();
    double Circumference();
    void CircleStats();   // This outputs the radius and center of the circle.
    Circles (float r);    // Constructor
    Circles();            // Default constructor
private:
    float  radius;
    int    center_x;
    int    center_y;
};

const double PI = 3.14;

//Client section

void main()
{
    Circles sphere(8);
    sphere.SetCenter(9,10);
    sphere.CircleStats();

    cout << "The area of the circle is " << sphere.AreaOf() << endl;
    cout << "The circumference of the circle is "
         << sphere.Circumference() << endl;

}
//___________________________________________________________________________
//Implementation section     Member function implementation

Circles::Circles()
{
   radius = 1;
}

Circles::Circles(float r)
// This procedure will take the radius of the circle from the user and place
// it in the appropriate member data.

{
    radius = r;
}

double Circles::AreaOf()
// This function will return of the area of the circle object that calls it.

{
    return PI * radius * radius;
}
```

continues

```
double Circles::Circumference()
// This member function will return the circumference of the circle object
// calling it.

{
    return 2 * PI * radius;
}

void Circles::CircleStats()
// This procedure prints out the radius and center coordinates of the circle
// object that calls it.

{
    cout << "The radius of the circle is " << radius << endl;
    cout << "The center of the circle is " << center_x
         << "   " << center_y << endl;
}

void Circles::SetCenter(int x, int y)
// This procedure will take the coordinates of the center of the circle from
// the user and place them in the appropriate member data.

{
    center_x = x;
    center_y = y;
}
```

Exercise 1: Alter the code so that setting the center of the circle is also done during the object declaration. This means that the constructors will also take care of this initialization. Make the default center at point 0,0 and keep the default radius as 1. Have sphere declared with initial values of 8 for the radius and 9,10 for the center. How does this affect existing functions and the code in the `main` section?

The following output should be produced.

```
The radius of the circle is 8
The center of the circle is 9 10
The area of the circle is 200.96
The circumference of the circle is 50.24
```

Exercise 2: As noted in the Pre-lab Reading Assignment, there can be several constructors as long as they have a different number of parameters. Alter the program so that the user will be allowed to enter just the radius or the radius and the center, or nothing at the time the object is declared. Whatever the user does NOT include (radius or center) must be initialized somewhere. There is no `SetRadius` function and there will no longer be a `SetCenter` function. You can continue to assume that the default radius is 1 and the default center is 0,0. Alter the client portion (`main`) so that it will reflect the changes by declaring an object `sphere1` by giving just the radius of 2 and using the default center, and `sphere2` by giving neither the radius nor the center (it uses all the default values). Be sure to print out the vital statistics for these new objects (area and circumference).

In addition to the output in Exercise 1 the following output should be included:

```
The radius of the circle is 2
The center of the circle is 0 0
The area of the circle is 12.56
The circumference of the circle is 12.56

The radius of the circle is 1
The center of the circle is 0 0
The area of the circle is 3.14
The circumference of the circle is 6.28
```

Exercise 3: Alter the program you generated in Exercise 2 so that the user will be allowed to enter nothing, or just the radius, or just the center or both the center and radius at the time the object is declared. Add to the client portion of the code an object called `sphere3` that when declared will have the center at point 15,16 and the default radius. Be sure to print out this new object's vital statistics (area and circumference).

In addition to the output in Exercise 1 and 2, the following output should be printed:

```
The radius of the circle is 1
The center of the circle is 15 16
The area of the circle is 3.14
The circumference of the circle is 6.28
```

Exercise 4: Add a destructor to the code. It should print the message **This concludes the Circles class** for each object when they are destroyed.

How many times is this printed? Why?

LAB 13.6 Constructors in Money Exchange

Retrieve program `monexc.cpp` from the Lab 13.2 folder. The code is as follows:

```
// This program converts dollars to foreign currency

// PLACE YOUR NAME HERE

#include <iostream.h>

const float DOLLAR_LIRE = 2048;           // Lire conversion per dollar
const float DOLLAR_FRANC = 6.937;         // Franc conversion per dollar

class Money
{

public:

    // Fill in the code to define three constructors: see Exercise 1

    float Lire() const;
        // This function returns Italian currency of current dollar and cent
        // amount
```

continues

```
        float Franc()const;
            // This function returns French currency of current dollar and cent
            // amount

        long Dollar() const;
            // This function returns dollar amount

        long Cents() const;
            // This function returns cents amount

        // Fill in the code to define a member function called PrintMoney that
        // will print out the value of dollars and cents to the screen.  The member
        // function should be a void function that prints the values within the
        // function itself

    private:

        long dollars;
        long cents;

    };

    void main()

    {

        // Fill in the code to declare an object called dough that initializes
        // dollars to 7 and cents to 80.

        // Fill in the code to declare an object called dough1 that has dollars and
        // cents initialized to the default values.

        // Fill in the code to declare an object called dough2 that has dollars
        // initialized to 10 and cents initialized to the default value.

        // Fill in the code to call PrintMoney for each of the objects declared
        // above

    }

    //  In the following space write the implementation for all the member
    //  functions of the class.  You did most of these in Lab 13.1
    //  Include the 3 constructors you defined above.
```

Exercise 1: Fill in the code that is asked for (in bold) in the above program. You are to define and implement three constructors. The default values for both `dollars` and `cents` is 0. The first constructor is the default constructor; the second requires the user to give both `dollars` and `cents` at declaration; the third requires the user to give `dollars` but not `cents`.

Exercise 2: Define and implement a destructor for the `Money` class that will print the following upon the object's destruction: "Money now out of space".

The output from the completion of both exercises should be:

```
The value of money is 7 dollars 80 cents
The value of money is 0 dollars 0 cents
The value of money is 10 dollars 0 cents

Money now out of space
Money now out of space
Money now out of space
```

LESSON 13.2B

LAB 13.7 Array of Objects

Retrieve program `Inventory.cpp` and `Inventory.dat` from the Lab 13.2 folder. The code is as follows:

```
// This program defines a class called Inventory that has itemnumber (which
// contains the id number of a product) and numofitem (which contains the
// quantity on hand of the corresponding product)as private data members.
// The program will read these values from a file and store them in an
// array of objects (of type Inventory).  It will then print these values
// to the screen.

// Example: Given the following data file:
//     986 8
//     432 24
// This program reads these values into an array of objects and prints
//      Item number 986 has 8 items in stock
//      Item number 432 has 24 items in stock
// to the screen

// PLACE YOUR NAME HERE

#include <iostream.h>
#include <fstream.h>

const NUMOFPROD = 10;    // This holds the number of products a store sells

class Inventory
{
public:

    void Inputid(int item);       // This puts item in the private data member
                                  // itemnumber of the object that calls it.
    void Inputamount(int num);    // This puts num in the private data member
                                  // numofitem of the object that calls it.
    void Display();               // This prints to the screen
                                  // the value of itemnumber and numofitem of the
                                  // object that calls it.
```

continues

```
private:

    int  itemnumber;            // This is an id number of the product
    int  numofitem;             // This is the number of items in stock

};

void main()
{

    ifstream infile;            // Input file to read values into array
    infile.open("Inventory.dat");

    // Fill in the code that declares an array of objects of class Inventory
    // called products. The array should be of size NUMOFPROD

    int pos;                    // loop counter
    int id;                     // variable holding the id number
    int total;                  // variable holding the total for each id number

    // Fill in the code that will read inventory numbers and number of items
    // from a file into the array of objects. There should be calls to both
    // Inputid and Inputamount member functions somewhere in this code.
    // Example: products[pos].Inputid(id); will be somewhere in this code

    // Fill in the code to print out the values (itemnumber and numofitem) for
    // each object in the array products.
    // This should be done by calling the member function Display within a loop

}

// Write the implementations for all the member functions of the class.
```

Exercise 1: Complete the program by giving the code explained in the commands in bold.

The data file is as follows:

```
986    8
432   24
132  100
123   89
329   50
503   30
783   78
822   32
233   56
322   74
```

The output should be as follows:

```
Item number 986 has 8 items in stock
Item number 432 has 24 items in stock
Item number 132 has100 items in stock
Item number 123 has 89 items in stock
Item number 329 has 50 items in stock
Item number 503 has 30 items in stock
Item number 783 has 78 items in stock
Item number 822 has 32 items in stock
Item number 233 has 56 items in stock
Item number 322 has 74 items in stock
```

LAB 13.8 Student Generated Code Assignment

Redo Lab 13.4 of Lesson Set 13.1 by including the following:

Replace the `Initial` member function by two constructors. One constructor is the default constructor that sets both `dollars` and `cents` to 0. The other constructor has 2 parameters that set `dollars` and `cents` to the indicated values.

Have the code generate two objects: `bank1` (which has its values set during declaration by the user) and `bank2` that uses the default constructor. Have the code input deposits and withdrawals for both `bank1` and `bank2`.

APPENDIX A

Visual C++ Environment

In this first lab, you will become acquainted with a software package that will allow you to create, compile and run C++ programs. Visual C++ is a product of Microsoft and is one component of Visual Studio (a complete set of programming language environments). The following steps walk you through most of the commands needed for this course.

I. Pre-lab Assignment

1. Be sure to read Lesson Set 1 Pre-lab Reading Assignment and do the Pre-lab Writing Assignment before coming to class.
2. Make sure you have a working password or some means of logging on to your particular computer system. Check with your instructor.

II. Visual C++ environment

1. Make sure you are logged on the system.
2. Select Start→Programs→Microsoft visual Studio 6.1→Microsoft Visual C++6.0 (Your particular system may have an alternative way of accessing Visual C++, check with your instructor).
3. After opening any source file, you must save it to your hard or zip drive before you will be able to compile and run it.
4. Once you are in the Visual C++ package you are ready to open or create files. All files are stored in a workspace.

III. Opening an Existing File

1. Click on the File Menu (top left corner) then select open.

   (The above step could be done by just clicking on the open folder icon).
2. You now see a window that says open. In the `Look in` box at the top there is a down arrow button. Selecting this helps you browse the drives to find where the files are.
3. Your instructor will tell you where and how to access the programs stored for this lab course, hereafter called the lesson folder.
4. We are going to open the first program. Double click on the `firstprog.cpp`.
5. This brings the file into the workspace; however, we must save it to our own disk space (hard drive or zip drive). Click on File→Save As→ and now use the browse feature to go to your hard drive and save the file there.
6. Your window should say something like `C:\firstprog.cpp`.

7. You can now compile this program by going to the **Build menu** and selecting **compile** (the first choice of the menu).
8. A message pops up that says the file requires an active project workspace. All Visual C++ files are stored in a workspace and must have a project created in that workspace. Just click **yes** to this message.
9. You should see some words at a bottom window that says no errors. When this occurs you are ready to run the program.
10. Click on build and then execute (it has a **red exclamation point**). This exclamation point may also be found somewhere on your screen.
11. Another message appears that asks if you want to build a file. Again just click on **yes.**
12. The program runs and shows an output display window. What is printed on the screen?
13. Hit any key and the message disappears.
14. After completing each lesson you MUST purge the Visual C++ workspace before opening a new source file. Do this by selecting File→Close Workspace. Select Yes when prompted about closing all document windows.
15. All programs brought in from the lesson folder must be saved to your hard or zip drive before compiling or running.
16. You can also create your own file from scratch by selecting File→New→Files then highlight C++ source file (if your are creating a program file) or text file (if you are creating a data file). On the right side, type in the name of the file in the box labeled file name, and determine where you want it saved by giving the drive in the box labeled location.
17. We do not explain at this point how to create a project. This is done for you when you compile a program you wrote or brought in from a lesson folder.
18. You have now completed Lesson 1A and are ready to begin Lesson 2A.

APPENDIX B

UNIX

In this first lab you will learn how to create, compile, and run C++ programs using the UNIX operating system.

I. Pre-lab Assignment

1. Be sure to read the Lesson Set 1 Pre-lab Reading Assignment and then do the Pre-lab Writing Assignment before coming to class.
2. Make sure you have a login name and a working password to logon to a UNIX system. Your instructor, along with the UNIX administrator, will help you obtain an account and learn how to login.

II. The UNIX Operating System

UNIX is a very powerful operating system. However, for some users it is not quite as easy to learn as the Windows environment. UNIX requires some extra knowledge on the part of the user. This appendix should help you learn some of the basics of this operating system. Nevertheless, you will most likely need additional information from your instructor or other sources to be a successful UNIX user. This appendix is a terse, but in no way complete, introduction to UNIX.

The UNIX system organizes files in a hierarchical structure. The top of this structure is called the root directory identified by `(/)`. Major directories are located under the root directory. For example, `/home` indicates the user home directory and `/bin` is the binary directory for executables. A directory called `programs` under `/home` is identified by `/home/programs`. The *path* `/home/programs/program1.cpp` shows the location of the C++ program `program1.cpp` in the `/programs` directory.

III. The UNIX Reference Manual

There is a UNIX reference manual that can be accessed during your UNIX session as long as your system has the `man` program. To get information regarding a particular command type `man` at the shell prompt followed by the desired command. For example,

```
man mkdir
```

provides you information about the `mkdir` (make directory) command. A list of useful commands that you may wish to explore as follows:

`cd`	changes directory
`cp`	copy file
`find`	find files
`grep`	locate text within a file
`ls`	displays contents of current directory

`man`	UNIX reference manual
`mkdir`	makes new directory
`mv`	move and rename a file
`pwd`	display name of current directory with path
`rm`	remove (delete) files
`rmdir`	removes directory
`vi`	begin a `vi` (visual interpreter) editor session

Of course, there are many other useful UNIX commands. To learn more about these commands and other important information, visit the UNIX resources at your library and local bookstore.

IV. Editing with `vi`

You can create and edit files during a UNIX session using the `vi` (visual interpreter) editor. First choose the desired subdirectory for your file. For example, if you are in the `/home` directory, enter `cd programs`. To start the editor enter `vi` at the shell prompt followed by the name of your file:

```
vi program1.cpp
```

There are two modes for `vi`: the text-entry mode and the command mode. First select the text-entry mode by typing the character **a** for append. Now you are ready to enter text.

Type the following:

```
#include <iostream.h>

void main()
{
     cout << "Welcome to the UNIX operating system" << endl;

}
```

You do not need to worry about what the code means at this point. Just make sure you can enter this using the `vi` editor. To exit the text-entry mode, press the `Escape` key. This will return you to the command mode. Editing commands can be entered from this mode.

A simple example of a `vi` editing command is the command to delete a character. If you want to delete a single character, move the cursor to the desired character and type **x**. To delete multiple successive characters, move the cursor to the first character and type the desired number followed by **x**. For example, **9x** deletes the next 9 characters. Other editing commands are the following:

`dw`	deletes a word
`dW`	deletes a word that contains punctuation
`ndw`	(or `ndW`) for deleting successive words
`i`	inserts new text (changes to text-entry mode)
`O`	opens a new line above (changes to text-entry mode)
`o`	opens a new line below (changes to text-entry mode)

For more information on these and other editing commands use the `man` command described above or some other resource on the `vi` editor.

To save the program you just entered, make sure you are in command mode and type **`:w`** and then press [ENTER]. Your work should now be saved in `program1.cpp`.

Finally, you will need to exit the `vi` editor before you can compile and run your program. To do this, type **`:q`** and then press [ENTER]. The shell prompt should now appear on your screen.

V. Compiling and Executing a C++ Program

Possible commands to compile your C++ program will depend on your particular system. Many UNIX systems have GNU C/C++ compile. In this case you may type

```
g++ program1.cpp
```

to compile your program. Assuming you typed everything correctly, there should be no errors and an executable will be created. A shell prompt will appear without any messages. Now type `ls` and then press [ENTER]. You should see a new file listed named `a.out`. To execute your program, type **`a.out`** and then press [ENTER]. The following should appear on your screen:

```
"Welcome to the UNIX operating system"
```

This indicates that you have executed your program. If `g++` does not successfully compile your program, see your instructor to find out the command necessary for your particular system.

Often, you will want to bring in an existing C++ program from a floppy disk or some other source. Your instructor will show you how to retrieve existing files from other locations.

VI. Logging Off

To log out upon completion of your UNIX session, type `logout` and then press [ENTER]. Make sure that you do not misspell `logout`—otherwise you may be leaving your files open to the next user!

Index